W0016879

Voicing Code in STEM

A Dialogical Imagination

Pratim Sengupta, Amanda C. Dickes, and Amy Voss Farris

The MIT Press
Cambridge, Massachusetts
London, England

The MIT Press would like to thank the anonymous peer reviewers who provided comments on drafts of this book. The generous work of academic experts is essential for establishing the authority and quality of our publications. We acknowledge with gratitude the contributions of these otherwise uncredited readers.

This book was set in Times New Roman by the authors. Printed and bound in the United States of America.

Library of Congress Cataloging-in-Publication Data
Names: Sengupta, Pratim, author. | Dickes, Amanda Catherine, author. | Voss Farris, Amy, author.
Title: Voicing code in STEM : a dialogical imagination / Pratim Sengupta, Amanda Dickes and Amy Voss Farris.
Description: First edition. | Cambridge, Massachusetts : The MIT Press, [2021] | Includes bibliographical references and index.
Identifiers: LCCN 2020015012 | ISBN 9780262045117 (hardcover)
Subjects: LCSH: Computer programming–Study and teaching. | Voice computing–Study and teaching.
Classification: LCC QA76.6 .S455 2021 | DDC 005.13–dc23
LC record available at https://lccn.loc.gov/2020015012

10 9 8 7 6 5 4 3 2 1

Voicing Code in STEM

Contents

List of Figures

List of Tables

Acknowledgments

The voices that make us are many: Dr. Pallavi Banerjee, for her generous, critical insights on gender, technology, society, and care work; Dr. Marie-Claire Shanahan, whose voice is ever present, especially in our imaginations of publicness; Dr. Miwa Takeuchi, for her Bakhtinesque brilliance and reflections on earlier drafts; Dr. Derek Beaulieu, for reorienting our relationship with the strangeness of language; Dr. Uri Wilensky, who planted seeds of many ideas in this book way back in the early 2000s; Dr. Rich Lehrer, for teaching us Science as Practice; Dr. Kevin Leander, for desiring more than design; Dr. Thomas Philip, for insightful comments on modeling and racialization; Dr. Jennifer Radoff, Dr. Erin Sohr, Dr. Ayush Gupta, and Dr. Andrew Elby for their critical insights and thoughtful commentaries; Dr. Andrew Hostetler and Dr. Ty Hollett for their partnership in our work on modeling ethnocentrism; Dr. Kara Krinks, Dylan Paré, Maryam Hachem, Basak Helvaci Ozacar, Erhan Ozacar, Stephanie Hladik, Marilú Lam Herrera, Apoorve Chokshi, Michael Cutler, Santanu Dutta and Megha Sanyal for their insights. Our utmost gratitude to Dr. Anna Pletnyova for inspiring us to engage with Todorov's work in ways that opened new worlds for us.

To the teachers and students who worked with us: thank you for opening up your classrooms and your lives to us. Your brilliance, generosity and kindness changed the ways we think about the possibilities for coding in classrooms. To Cherifa McDowell, Emily Bryant Harper and Leah Keys Embry: working with you transformed us in the deepest possible sense.

To our families: Pallavi, Daniel, and our heterogeneous little ones—Ari, Chulbul, Ginger, Shep, Trapper, and those that cannot be named here—you are our lives. Thank you for making us who we are.

To Susan Buckley, the most amazing and supportive editor anyone could hope for: Thank you for everything.

—*The Authors*

1 Beyond Technocentrism: Coding as Experience

1.1 Setting the Stage

The last few years have seen a resurgence of interest in making everyone coding literate. Coding academies have popped up faster than one can count, public libraries and schools now have makerspaces, and there is interest in helping even preschoolers learn to code.[1] Political leaders appear in national advertisements proclaiming the importance of coding, urging their citizens—especially the younger ones—to become coders.[2] Coding is no longer confined to the ivory tower of the academy: in North America, more than 40 US states have adopted versions of K–12 Computer Science Standards or Next Generation Science Standards, and several Canadian provinces include computer science in their curricula.[3]

The popularized premise behind "coding for all" can be stated in simple terms, as follows.[4] By learning programming or coding (we use these terms interchangeably throughout the book) and by learning to think like a computer or a computational scientist, students will become better problem solvers, designers, and innovators. At the same time, it will make students jobs-ready. Teaching everyone how to code (that is, "write" computer programs) and how to *think* computationally have now become major foci for educational researchers as well as national and international funding agencies that support educational research and practice. In particular, the increasing focus on *computational thinking*,[5] both nationally and internationally, has resulted in a push to introduce computer science as part of K–12 curricula[6] as well as to integrate computing and coding into the science, math, and engineering curricula in K–12 classrooms.[7]

The *essence* of technology, however, is not the technology itself.[8] Throughout this book, we will explore the meaning of this dictum in depth; here, we provide a cursory reflection on what this means for coding, particularly in the context of education. Coding—commonly understood as the act of authoring, creating, or modifying computer programs or code—has long been argued to be a powerful form of learning experience. In his pioneering book *Mindstorms*, Seymour Papert argued that coding allows even young children to use the computer as a *malleable* tool that can take the shape of their imagination. Used in this way, coding can emancipate the learner from the restraints of otherwise cognitively limiting curricular mandates. For example, learning about speed and motion can be accomplished by learning to "paint,"[9] and learning mathematics can be reframed as game design.[10] This view of knowing and learning has been positioned as constructionism, in which central to learning are acts of constructing artifacts that are rich in terms of disciplinary knowledge, valued socially, and at the same time, personally meaningful.[11] Constructionism stands in opposition to stereotypical images of classrooms and schools that rely primarily on didactic teaching, which Papert famously termed "instructionism."[12] What counts as disciplinary knowledge can then expand and deepen simultaneously, when personal narratives can be seamlessly woven with disciplinary forms of knowing.[13]

Our work, however, arises from the concern that despite a wide recognition of computing as a potentially empowering experience for learners, the complexity of *experiences* of coding—both in general and more specifically in K–12 science and math classrooms—has not been studied in sufficient richness and depth. This is not to say that researchers have not studied how students can learn to code. In fact, the growing number of handbooks on educational computing would indicate the reverse, and both cognitivism and sociocultural theories of learning have shaped these efforts.[14] For example, a prominent strand of research on educational computing has historically focused on "misconceptions" and challenges faced by individual students in learning programming.[15] Over time, the field has also broadened its focus to include anthropological[16] and social[17] dimensions of learning to code. However, despite these advances, in many investigations of coding, the emphasis remains on justifying the design decisions taken by the researcher-designers of the technological infrastructure, and the "charisma" of computing machines and programming languages are positioned at the center.[18] This limits the scope of what should count as helpful or legitimate experiences of coding *in context*, despite a substantial expansion of the range of technological artifacts and environments in which coding takes place over the past few decades.

We dive deeper into Papert's critique of such *technocentric*[19] approaches in the following section (and throughout the book), as it undergirds an important foundation of our work. Papert positioned technocentrism as the fallacy of referring all questions about technology (including questions about the experience of technology) solely to the technology itself. It is also important here to remind ourselves of early warnings by Sherin, diSessa and colleagues[20] and Guzdial,[21] who noted that simply introducing a general-purpose programming language in science may make teaching and learning science even more challenging. With these warnings kept in mind, this book offers a detailed examination of complex experiences of the integration of computational modeling and programming with K–12 science and math. It advances the epistemological study of computing beyond the learner-centered paradigm, and argues for shifting our focus away from *computational artifacts* to *computational utterances*, and from *ownership* to *authoring*. Against the current state of a field that has been greatly influenced by notions of *thinginess*,[22] we offer *language* as the central metaphor for code and coding, particularly in K–12 STEM classrooms and contexts. This is not simply the story of redesigning a programming language so that learners and teachers can use them effectively or easily: it is the story of how computing becomes *language in use* in K–12 disciplinary contexts and classrooms.

1.2   A Two-Part Argument against Technocentrism

In essence, this book offers a two-part argument. The first part is a general concern about the nature of coding that goes beyond the specific issues of curricular integration of coding. We believe that educational researchers need to consider coding as a far more complex experience than writing code, remixing or manipulating programming commands, or constructing physical computing machines. We argue that as the emphasis shifts from studying whether learners accomplish the desired computational outcomes to how they learn to *voice* code as part of their contextualized work, new elements of human experience beyond device-level engagements come to light. The interplay between computational abstractions and the fundamentally interpretive nature of human experience must be studied carefully. Careful attention to these elements of experience allows us to design better tools and curricula for students, and enables us to see computing as experience beyond the mastery of symbolic power.

In the context of educational computing, this calls for paying attention to the complexity of the learning experience beyond a narrow focus on formal representations were made as early as in the 1970s.[23] Our project here is a renewal of this call, albeit with a significant epistemological relocation, as evident in our explicit commitment to the *heterogeneity of language* as the central metaphor for computing.[24] This entails not only conceptualizing coding beyond formal computational representations, but also shifting our focus beyond and, as necessary, away from *device-level engagement* with the computer. We borrow this phrase from Daniela Rosner, whose critique of Herbert Simon's influence in the field of human-computer interaction was similar: " . . . Simon reinforced a model of computer interaction that limited analysis to device-level engagements, reducing the world to well-structured problems."[25] In our case, the word *device* refers to the programming language in use, as well as programmable media such as microcontrollers, sensors, and so forth that typically serve as the *technologies* for coding. We see device-level engagement as a central focus of technocentric approaches to computing in education, which reduces the experience of coding to computer programming or working with programmable media.

In addition, viewing coding as a heterogeneous form of voicing should compel us to pay explicit attention to *unsilencing* voices and forms of discourse that are often left out of educational computing and hence to our commitment to critical phenomenology. The critical phenomenological position argues against a universalist notion of experience, because what counts as "experience" in a discipline is only the experience of a few.[26] A case in point here is Morgan Ames's critique of the One Laptop Per Child (OLPC) project. Ames argues that the OLPC project was modeled after an imagined middle-class, White user, despite stark differences with the contexts in which the project was enacted.[27] This is also an essential dimension for untethering our understanding of code and coding from device-level engagement, because it brings into focus experiences that historically have not been seen as central to acts of knowing in the context of computing.

The second part of our argument is deeply contextualized, as it deals with some of the peculiarities of the classroom as the place where coding is experienced. The classroom is not merely a physical location but is also a community that is distinct from the online and virtual communities that Kafai and Burke refer to in their book *Connected Code*. Learning in classrooms is not independent of teachers, who have to balance several complex forces, including curricular mandates that have specific disciplinary and performative commitments. The complexity of teaching in classroom contexts that include coding has barely been investigated, even though early studies that tried to focus on teaching reported that the curricular integration of coding in science is challenging.[28] Of particular note is the historically oppositional stance within constructionism to schooling and instruction.[29] The "avoidance of direct assistance"[30] is an epistemological position that is deeply rooted in the constructionist literature that has positioned a range of technologies from programming languages to makerspaces as "tools for emancipation."[31] In addition, the complexity of teacher-researcher partnerships and collaborations has not received any significant attention in educational computing.[32] Our concern is that leaving these forms of experience out of our focus in educational computing research can edify essentialist images of device-level engagement as the seat of learning. Technological determinism—synonymous with technocentrism—is the over-reliance on these images as images of educational computing.

In order to illustrate images of heterogeneity in coding, we ask the following question: How does code *become* experience in K–12 math and science classrooms? In asking this question, our goal is to go beyond technocentrism and technological determinism. Papert's critique[33] of studies that were conducted to "evaluate" Logo—both positively[34] and negatively[35]—is particularly poignant in this regard. Technocentrism is embodied in these studies in the form of a "treatment" model: if children are provided with particular tools, they will learn X.[36] Whereas a particular representational format—for example, drawing a circle using a Logo Turtle—may be an easier way to learn geometry from a cognitivist perspective, the actual experience of learning in the classroom depends on several factors beyond the availability of the computer and the programming language such as the quality of teaching, institutional supports for technology integration and teacher preparation, the historical positioning of students in their relationships with the institution and the disciplines, and so forth. The lack of focus on these issues can lead to erroneous interpretation of both positive and negative results, because the difference between an effective and ineffective use of Logo may be influenced significantly by these other factors, as well as by Logo itself.

But Papert also specifically alerted us to the danger of technocentric approaches in the context of interpreting *negative* results. By focusing solely or primarily on the tool or the technological artifacts, such approaches often fail to recognize "social structures and cultures that existed before the computer"[37] that continue to shape the experience of computing. The work of Jane Margolis and her colleagues provides an illustration of this problem. Their careful analysis shows that cultural and institutional factors are often ignored by practitioners and researchers in interpreting why high-school students of color perform worse in computing-intensive curricula, which then positions racially marginalized students as academically deficient. Herein lies a truly grave danger of technocentrism. We revisit this issue in more detail in section 1.3.2.

On a more introspective note, let us consider the invention of new technologies for learning—such as a new programming language or an online platform for helping students learn to code. Such inventions have been an important focus for us as educational computing researchers. Inventing such technologies may not be inherently technocentric; but, in a technocentric frame, even research on learning using these technologies is largely geared toward evaluating the efficacy of the technology itself. Add to this the recent emergence of data mining and machine learning as new and powerful analytic approaches in educational research, which often cause the experience of coding to be inferred primarily—in many cases, exclusively—from the code created by the students rather than from observations of their experiences.[38] Technocentrism also manifests itself by framing teachers in deficit lenses or in disempowered positions in our classroom studies, in which the technological and pedagogical decisions have largely been taken by the researchers, leaving teachers with limited agency to appropriate the tool and curriculum. The examples are simply too many to list.

In fact, we must admit here that some of our own previous work has been subject to these issues. So, as we (the authors) reflected about this, we realized that this is not surprising, because for us as technologists and tech designers, technology *is* a significant element of our experience. This makes us prone to view studies of coding as contexts that can justify our design decisions rather than as opportunities to study the experience of coding beyond the students' work on the computer. This in turn leads us to select only the bits of the experiences of the teachers and the students that speak directly to some of the assumptions or theories inherent in our technology and curricular design decisions. So, we end up being boxed within the technocentric frame by using images *within* technology to understand the experience of technology. The analysis of learners' voices and actions that usually unfold *outside* the computing media has received less attention. And in the classroom, as we show throughout the book, careful attention to these elements of experience can reveal important insights about the uncertainties and subjectivities that constitute the experience of coding.

We argue then for looking at coding as a particular form of experience: the interplay between artificial and natural languages. The genre of *artificial languages* includes programming languages and modeling toolkits, and we use the phrase *natural languages* broadly, to include verbal, gestural, and symbolic interactions and exchanges, which are also shaped by our historical positionings. Our project here is an epistemological inquiry, and Bakhtin's dialogic and heterogeneous imagination of language serves as the central underlying metaphor *for* our inquiry.[39] We explore Bakhtin's work in detail in the following chapter, in which we also outline a framework for "seeing" and "hearing" coding as heterogeneous and heteroglossic discourse. Enframing Bakhtinian multivoicedness within a critical phenomenological approach, we argue, can help us counter technocentrism in educational computing. However, before we engage in a dialogue with Bakhtin's work, let us first consider more deeply what we mean by critical phenomenology and what such an approach can offer to the study of coding, particularly in K–12 STEM contexts and classrooms.

1.3 Toward a Critical Phenomenology of Coding

The *phenomenological* dimension of our approach involves refocusing our attention to the fundamentally complex nature of human experiences that are involved in coding and learning to code. The *critical* dimension involves learning to recognize voices that historically have received less attention and have been valued less in computing education, and more broadly, in technoscience and public education. In this section, we first present a phenomenological counterpoint to technocentrism by arguing to move beyond the emphasis on symbolic generalizabilty. We then introduce some of the key concerns from more critical and historical perspectives that further challenge the technocentric paradigm by alerting us to the dangers of relying on theoretical formulations that may without our awareness perpetuate certain forms of silencing of marginalized experiences.

1.3.1 The Phenomenological Turn: Beyond Symbolic Generalizability

At the heart of our phenomenological agenda is the notion of *sense experience*. Merleau-Ponty defined sense experience as "that vital communication with the world which makes it present as a familiar setting of our life."[40] Through this dialectical form of communication with the world, perception plays the role of "pointing out," performing an act of voicing (that is, giving voice) to the texture of experience in its very act of happening.[41] Merleau-Ponty argues that this involves a form of radical reflection, because the phenomenal field resists being explicit to our experience, hindered due to the *sphere of givenness*,[42] that is, categories that are simply bestowed upon our senses through social and institutional forces that most directly shape our language and thus our communication with the world. It is a form of reflection that turns over forms of speech that are given to us, in order to reveal our "originary" sense of the phenomenal field. Citing Merleau-Ponty, the phenomenologist Laura McMahon paints this form of reflection beautifully: "It is in the midst of this 'strange and paradoxical' world of perception that the indeterminate, open-ended identities of perceived objects and perceiving subjects are simultaneously born."[43] McMahon compares this image to that of a city dweller who is "born and reborn in their acts of perceptually "pointing out" the subtle aspects of what is there to be seen" in their city of dwelling, thus an act of continual place-remaking.

Our phenomenological project relies on acknowledging the inseparable relationship between knowledge and knowing, ideas and actions, and actions and interactions. In doing so, our intention is to necessarily blur the delineation between ideas, bodies, and the world; algorithms and hardware; and artificial and natural languages—inasmuch as they constitute our experiences of computing, particularly in STEM disciplines and K–12 STEM contexts. To blur the delineation is not to deny the existence and importance of each. Rather, our goal is to bring into focus the inseparability of each of these aspects *of experience* from the others. To understand how students and teachers may come to adopt coding as a language in their science classrooms, we must come to see all of these aspects of human experience as central to our efforts.

At first glance, such a project might seem antithetical. Shouldn't coding and studies of coding put code at its center? One could legitimately argue that *computational thinking*, the phrase that has indeed come to shape much of our current efforts in computing education, positions computing as the "automation of abstractions."[44] This positions code and the computer at the center of accounts of what should count as computing. But, one could also interpret Wing's push for *computational thinking* (rather than *computing*) as an important phenomenological opening, in the sense that the emphasis on "thinking" can potentially bring to light some element of the "sense experience"[45] of computing professionals as they deal with computational abstractions. And yet, our phenomenological concern here is that we need to move beyond the veneer of computational abstractions and computational thinking in order to develop a richer accounting of the heterogeneity in the experiences of coding and computing in STEM classrooms and disciplinary contexts. Narrow and technocentric adoptions of these constructs position them as *second-order perceptions*,[46] which can severely limit the scope of heterogeneity in our visions of computing by centering the reproduction and/or regurgitation of computational abstractions in students' computer programs as the only (or primary) evidence of learning. For example, most of the studies of children learning to code as reviewed by Grover and Pea[47] fall within this category. Rather than revealing the complexity of experiences, second-order perceptions serve as *crypto-mechanisms*,[48] that is, simple mechanisms that hide more complex forms of experiences. By focusing primarily on these constructs, we tend to overlook the significantly more complex and uncertain dimensions of the experience of code and coding, thus potentially making such experiences irrelevant or ancillary to the study of learning to code.

In order to understand our concerns a bit more clearly, let us take a closer look at Wing's own elaboration and description of computational thinking. Wing positioned computational thinking as an analytic approach for

> solving problems, designing systems, and understanding human behavior, by drawing on the concepts fundamental to computer science.[49]

Positioning computer science as the automation of abstractions,[50] Wing argued

> the most important and high-level thought process in computational thinking is the abstraction process. Abstraction is used in defining patterns, generalizing from specific instances, and parameterization. It is used to let one object stand for many. It is used to capture essential properties common to a set of objects while hiding irrelevant distinctions among them.[51]

Wing's conceptualization of abstraction emphasizes the notion of generalization: Abstractions are generalizable, *symbolic* representations that can be *applied* in multiple situations or contexts. An algorithm, Wing[52] noted, is an example of an abstraction because it is a generalizable representation of a step-by-step procedure that generates a desired output based on a specific type of input. Similarly, a programming language can also be considered as a layered abstraction of a set of *strings* that are data structures in themselves. Each string, when interpreted by the computer, effects some form of computation. The programming language that we have used for the empirical work reported in this book—ViMAP—is an even more complex form of abstraction, because it is a combination of two different programming languages: NetLogo and Java. Noting that "abstractions are the 'mental' tools of computing,"[53] Wing further claimed that when scientists and engineers use "cleverer or more sophisticated abstractions," they may be enabled to "analyse their systems on a scale orders of magnitude greater than they are able to handle today."[54]

Wing's articulation of computational abstractions as forms of symbolic generalizations thus bears similarities to the position of philosopher John Locke, who viewed abstraction as a mental process. Locke proposed two types of ideas: particular and general. Particular ideas are constrained to specific contexts in space and time. General ideas are free from such restraints and thus can be applied to many different situations. In Locke's view, abstraction is the process in which "ideas taken from particular beings become general representatives of all of the same kind."[55]

In a similar vein, an interesting facet of generalizability also evident in Wing's claim is that computational thinking does not necessarily entail the use of a computer and neither should it be equated with programming. Wing further noted that thinking computationally can be often ubiquitous in our everyday experiences:

Consider these everyday examples: When your daughter goes to school in the morning, she puts in her backpack the things she needs for the day; that's prefetching and caching. When your son loses his mittens, you suggest he retrace his steps; that's backtracking. At what point do you stop renting skis and buy yourself a pair?; that's online algorithms. Which line do you stand in at the supermarket?; that's performance modeling for multi-server systems. Why does your telephone still work during a power outage?; that's independence of failure and redundancy in design.[56]

This is a powerful quote, because it implies that even apparently mundane, ubiquitous experiences can become the realm of computational thinking. In fact, a not-so-hidden implication is that some of the more complex elements of our lived experiences can be dealt with more easily using computational thinking. But a phenomenological look offers a more intricate image of the experience of thinking computationally in such contexts, because it is not simply a matter of "applying" computational thinking in different situations. On the contrary, the relationship between abstractions and contextualization is a deeply intertwined one, as we explain next.

Let us examine carefully the analogy central in Wing's quote: choosing which line to wait in, and the computational abstraction of performance modeling for multiserver systems. The latter is at once a problem in computer science, as well as in operations research—the field in which Herbert Simon, the pioneering computer scientist, made significant discoveries about computational intelligence. One could interpret Wing's argument in this analogy as follows: elements of the human experience of selecting a line to wait in can be understood more deeply, and perhaps even improved upon, if we are able to recast this problem in terms of computational abstractions that computer scientists use to study and improve computer networks. Our goal here is not to deny the value of this form of reframing of an everyday experience in terms of computational abstractions. Instead, we seek to highlight the elements of the experience—the *work* involved in doing such reframing—that gets left out when the *essence* of such reframing is reduced to the resultant abstractions. One could plausibly argue that the imagination leading up to the articulation of the analogy is one in which contextuality and computational abstractions are deeply intertwined and also, quite possibly, shaped by previous professional and/or pedagogical experiences. This is the *phenomenological turn* we are arguing: a greater accounting of the *experiences* through which abstractions become evident in context.

The professional practice of computation is highly contextualized.[57] As the noted computer scientist Douglas Schmidt has pointed out, the professional practice of software engineering necessarily involves modeling and design *grounded in contexts of application.* A phenomenological perspective would indicate that it is through engaging in these practices that computational abstractions become usable software through being progressively recontextualized, and this becomes evident as we take a closer look at modeling, one of the central practices in software engineering. Modeling involves deciding what should be highlighted and represented, and complementarily, what needs to be ignored, depending on the contexts of application. Here's an example of modeling *in context*: computer scientists working on the problem of performance modeling in multiserver systems acknowledge the centrality of a deep understanding of the complexity of the application characteristics for designing their computational models. These characteristics include resource needs of different components, intercomponent communication delays, and so on, without attention to which their models would fail to work effectively. At the same time, computer scientists also acknowledge which elements of the context should be ignored in their model, for example, the operating system used by the server.[58] The emphasis on contextualization is unmistakable here, as modeling positions computing as design.

The practice of designing usable software also involves the choice of computational abstractions and programming language appropriate for the task at hand. Increasingly, we are seeing the emergence of *agile* computing—a key characteristic of which is the strong dependence of the *form* of computing on the *context* of the application.[59] The choice of abstractions, as Schmidt points out, is based fundamentally on thinking contextually about specific problems rather than dealing with all the complexities of the underlying computing environment (e.g., CPU, memory, and network devices). Note that this is also evident in Wing's articulation of the importance of the material and physical constraints underlying the information-processing agent and its operating environment.[60]

This suggests that we may be in a linguistic trap: although Wing defined computational thinking as *thought processes*,[61] in reality, the experience of thinking computationally, especially in STEM disciplines, is intricately interwoven with the disciplinary context at hand. While the choice of appropriate computational abstractions can greatly advance scientific and technological inquiry, the contextualization of these abstractions is central to understanding how we experience computing and computational thinking. Our point here is not to deny that computational thinking involves either *thinking* or *generalizability*. Instead, our argument is that if we are to understand how computational thinking is experienced by students and teachers, we must simultaneously broaden and deepen our inquiry into their experiences beyond rather thin notions of "thinking" and device-level representations of "computational abstractions." In a stark contrast to folding experiences within the technological interfaces—the essential crisis of technocentrism—we seek to provide accounts that illustrate how computational abstractions become meaningful through iterative and complex forms of *recontextualization*. So, here's a fundamental insight from a phenomenological perspective: the experience of computational abstractions is a form of recontextualization.

Our book reveals how computational representations and abstractions become meaningful through being interpreted and reinterpreted both in light of the lived experiences of the students and the teachers and in terms of the disciplinary practices that are central (in our context) in non-CS, STEM disciplines. In the empirical work we report in this book, both these forms of recontextualization become explicit as students and teachers engage in *modeling*, which has been recognized as the "language" of science and holds a central place in K–12 science education.[62] We take a deeper look into the relationships between modeling and coding in the following chapter. In Merleau-Ponty's terms, our goal here is to present a "first-order perception" of what becomes of code and coding in math and science curricular and classroom contexts as students and teachers engage in modeling. We show that in doing so, a much more complex "phenomenal field" (Merleau-Ponty, 1962), that is, experiences and lifeworlds within which code becomes meaningful, comes into view. What constitutes such a phenomenal field itself can (and should) be a matter of debate, and as such, in this book, we present some of the key elements that might constitute such a field. What we present, nonetheless, is a rich tapestry, including materiality (chapters 4 and 6), perspectival work (chapter 3), engaging in difficult conversations about race and ethnocentrism using computational simulations (chapter 5), and carework (chapter 7).

1.3.2   Critical and Historical Concerns

The critical historical context in which our work is situated rests on prominent feminist and post-structuralist critiques of computing and technoscience, and a careful attention to the racialized history of abstractions in public education. Feminist critiques of computing and technoscience have identified technological determinism[63] and masculinist notions of a "pure" discipline[64] as deterrents for equity in cultures of computing. Central to these critiques is the repositioning and reimagining of the human-machine boundary by challenging the orthodoxy of a cognitivist vision of computing, which reinforces and restricts studies of computer-human interaction to device-level engagements.[65] The ambiguity and historicity of human experiences with computing remain largely hidden from view in such studies, and with them, the lifeworlds of the people engaged in computing.

At stake here is our perception of what computing is and what it should look like in the classroom. Our concern is that this form of thinking perpetuates a *universalizing*[66] aim of computing as a dominant discourse, especially in K–12 education. Haraway termed this the "hegemony" of technoscience. The power of computing, as we indicated in the previous section, lies in its offering a generative core that can take on malleable forms—for example, a generalizable algorithm that can explain many different situations and offer many different solutions in different contexts. Rosner reminds us of a quote by Susan Leigh Star that is quite apt here: "Power is about whose metaphor brings worlds together."[67] Our critical historical concerns here are grounded in the perspective that the way in which we negotiate our relationship with this power in pedagogical contexts must also be informed by the troubled histories that underlie some of the constructs central to the *pure* discipline of computing (e.g., *thinking* and *abstractions*). We situate these histories in the realm of public education in the US and in critical ethnographic analyses of technological design in practice. The implication of this kind of critical historical framing is the need for disciplinary heterogeneity that goes beyond reshaping disciplinary boundaries, also bringing into account voices from the margins that typically do not fall within the purview of computational thinking.

A puritanical view of computing adheres primarily to a cognitivist agenda, and it is not uncommon to draw parallels between computing and the mind. For example, Wing noted that computational thinking "shares with scientific thinking in the general ways in which we might approach understanding computability, intelligence, the mind and human behaviour."[68] And theories of computing have been developed empirically by studying and modeling "intelligence" using computational abstractions. Such an example is Newell and Simon's theory of computational intelligence, which as the critical scholar Alison Adam notes, leaves out any considerations of race, class, and gender, treating "intelligence" as a universal category.[69] On one hand, it seems tempting to argue that this is, in fact, an egalitarian position that assumes everyone is capable of intelligence. However, a historical look at how intelligence and its inevitable computational twin—abstractions—have been positioned by scholars in public education in the US would reveal a troubling image. The overtly racist history of *abstractions* in public education in the US[70] should serve as a cautionary note against adopting such positions unproblematically. Margolis reminds us that as early as 1923, immigrants, Mexican, and Black people had been represented in the intelligence-testing literature as people who "cannot master abstractions, but they can often be made efficient workers."[71]

Noteworthy in Terman's horrific statement is also the intertwined nature of ideologies of racial differences and the mind-body divide. As Mike Rose[72] reminds us, the distinction between the mind and the body has also been positioned as the distinction between abstractions as the realm of higher-order thinking, and work (including physical work) as the realm of the more commonplace forms of functioning. This distinction is also tied to class distinctions in society. Rose's work, as well as recent scholarship on embodied cognition, calls this distinction into question. Rose, for example, eloquently argued that both physical labor and vocational education include significant forms of higher-order thinking. Similarly, through a careful analysis of the interpretive and communicative work involved in car repair by car mechanics, Streeck has shown that such forms of physical labor rely on the creative use of abstractions in the form of complex gestural communication.[73]

The work of Margolis and Rose has collectively shown that racial disparities in computing education and the abstraction-vocation divide are both alive and well in US public education and, further, implies that these issues are interlinked. Margolis and her colleagues have been studying the question of why African American and Latino students in the US have predominantly been left out of the folds of educational computing. Their work orients our attention to "virtual segregation," which they defined as an "insidious phenomenon that occurs when we are led to believe that we are moving toward equality, and pretend that everyone has a chance and a choice."[74] Margolis argued that this perception stands in contrast to the reality in which the playing fields are immensely uneven because of both structural inequalities (e.g., lack of good computing curricula in predominantly African American and Latino schools) and widely held beliefs about the low expectations from students of color. She reminds us of the famous words of mathematics educator Robert Moses that the knowledge gap in math, science, and technology is a civil rights issue for the 21st century, as this could turn students of color into the "designated serfs of the information age."[75] Margolis also argued that deficit-perspectives held by teachers regarding abilities of Black and Latino students actually result in the low performance of these students in computing classes, and more recent analysis of nationwide educational achievement data shows that teacher bias and other systemic obstacles hinder overall academic progress for Black students in the US.[76]

It is therefore imperative for us to think carefully about *where* to situate computational abstractions. The predominant adoption of Wing's notions of computational thinking and computational abstractions locates them *in the mind* in the form of thought processes.[77] This unwittingly reifies the divide between vocational and intellectual work and furthers conceptualizations of technology that fail to consider the broader, infrastructural view beyond the child-computer dyad. As Margolis and colleagues have argued, this eventually leads us to erroneous conclusions about the *ability* of children, especially from marginalized groups whose struggles to overcome significant historical and systemic obstacles often remain invisible in their classroom work.

In embracing more expansive enframings of code and coding, we examine closely—albeit in STEM curricular contexts—the breakdowns in human-code communication, the necessarily distributed nature of experiences beyond the individual, and the material and non-computational forms necessary for enlivening symbolic code. By doing so, we challenge the dominant paradigm in which studies of computing in education focus primarily on universalist and individualist forms of computational abstractions. In contrast, we argue that we should also pay attention to the possibilities for *différance*[78]—that is, both difference and deferral—of meaning that might emerge from "aberrations in the narratives at hand,"[79] rather than amplifying the desire for immediate control.

1.4 Epilogue: Notes on Situatedness, Epistemology, and Methodology

We began this chapter with a critique of technocentrism, arguing for a shift from viewing coding only as an engagement with a programming language to a much more heterogeneous, contextualized, and historically situated form of *experience*. But it is also important to note that early epistemological positionings of the Logo Turtle were grounded in feminist scholarship that valued intimacy, affect, and embodied ways of knowing as central to learning to code.[80] To know, from a feminist standpoint, is to relate; to gain deeper understanding is to grow-in-connection.[81] In *Mindstorms*, Papert noted that computational agents act as *transitional objects*,[82] that is, mediational objects which learners can use as proximal projections of their lived experiences. The Turtle's movements on the computer screen can be interpreted through embodied actions that even young children are able to perform, and thus the Turtle becomes body-syntonic. The experience of programming thus becomes a conversational one, in which writing a program becomes an experience of teaching the Turtle how to carry out an action. Such approaches to computing allow for epistemological pluralism by making space for different ways of knowing (e.g., both planning and tinkering) in the computing classroom.[83] As Wilensky[84] argued, this is indeed an image of *concreteness* in which learning computing becomes an experience of developing a progressively closer relationship with the computational agent and the programming language. Such a situated and relational image of programming is similar to Fox-Keller's sketches of the Nobel Laureate Dr. Barbara McClintock's experiences of *feeling for the organism* as a way of developing new insights in biology.[85]

The programming languages we have used for the empirical work in this book are all agent-based languages derived from Logo. This, however, is a double-edged sword. On one hand, there is now a plethora of studies that have demonstrated how children can indeed take advantage of the situatedness that is inherent in thinking and programming with computational agents and develop deep conceptual understandings in K–12 science and mathematics. On the other hand, an unproblematic valorization of any programming language as sufficiently situated is also something we should guard against. Solely attributing situatedness to a particular form of programming can limit our attention to the child-computer dyad as the *site* where computing is taking place. In such images, learning programming is recast as a child developing control over the *protean* (to use Papert's term) Turtle, thereby reducing the fundamentally social experience of learning a language to a "self-directed" and individualistic one.[86] Vossoughi, Escudé, and Hooper[87] have further argued that "overreliance on child-centered pedagogies that emphasize the avoidance of direct assistance overlooks the powerful role intentional teaching can play in challenging deficit ideologies and cultivating substantive experiences of intellectual dignity."[88]

Ames also reminds us of studies conducted by researchers in the US and the UK, which showed that learning gains with Logo would often vanish when researchers were not present in the classroom.[89] While Papert would argue that these studies were carried out in the technocentric paradigm, our focus here is on the finding that teaching is an essential component of the experience of learning coding in classrooms. Our own research in classrooms also reveals a similar picture: while children may be able to understand some of the more challenging scientific and mathematical ideas using computational modeling and coding, they also experience significant challenges in their learning experiences. These challenges arise from linguistic uncertainty, conceptual difficulties related to learning programming, and difficulties with interpreting scientific and mathematical ideas and representational practices. Furthermore, we found the role of teaching to be key in scaffolding students through these experiences.[90]

This then raises another concern: researcher-teacher relationships, which bring with them complex issues of power and agency, especially from the perspective of the teachers who are often positioned in deficit lenses. Teachers in

STEM disciplines typically do not bring in prior experience with computing and programming in the same way that the computing education researchers do. This power differential is further aggravated by the broader history of positioning teachers *not* as experts when they work with researchers, especially in science education.[91] So, this suggests that in order to understand the social and curricular contexts of the classroom, we need to *listen* to how teachers are engaging with programming and computing *from their perspectives*—an essential situatedness that is largely missing in the field of computing education.

Finally, another important dimension of situatedness comes from the disciplinary contextualization of code in K–12 classrooms. This is both an epistemological and a methodological concern for us. Both these concerns, in turn, are deeply intertwined with how we conceptualize disciplinary cultures in which students and teachers are (and should be) situated. By framing coding as language and discourse beyond the myopia of the child-computer dyad, our epistemological objective is to bring to focus the ambiguity and interpretive nature inherent in computational discourse in the way it plays out in the social world of the classroom. Our methods of observation and analysis should also be aligned with disciplinary practices in science. To this end, it is important to note that our positioning of computing and coding as discourse is also juxtaposed with the reframing of coding as *modeling* in STEM disciplines.[92]

This in turn allows us to draw upon a rich body of literature in science studies and science education that have long valued uncertainty, interpretive experiences, and multimodality as central to doing and learning science, even when the result of this work may have the appearance of certainty.[93] Our choice of Bakhtinian heterogeneity as the lens to understand experiences of students and teachers in our classrooms is thus grounded in the epistemological positioning of heterogeneity[94] and uncertainty[95] as central to modeling in science and STEM classrooms. We explore this reflexivity in more detail in the next chapter (chapter 2), and the empirical chapters illustrate heterogeneity *in action* in different ways (chapters 3–7). Our studies also show how disciplinary purity and curricular homogeneity must necessarily be at risk when the multi-disciplinarity inherent in designing code in STEM classrooms is not reduced to the reproduction of computational abstractions. In such contexts, code talk can become race talk as computational algorithms become recontextualized through perspectival projections of lived experiences on computational agents (chapter 5), and historically meaningful, religious symbols can offer initial sites for powerful computational work for marginalized students (chapter 7).

Notes

1. See, for example, a recent NPR story on coding in preschool:
http://www.npr.org/sections/ed/2015/09/18/441122285/learning-to-code-in-preschool.
2. President Obama's public address on coding:
https://www.youtube.com/watch?v=6XvmhE1J9PY.
3. See, for example, a recent K-4 curriculum for Alberta, Canada:
https://www.alberta.ca/release.cfm?xID=60779ADBF92D9-BC17-6738-EED797C3813AC9F0.
4. See the US Government's official policy briefing on CS for All:
https://obamawhitehouse.archives.gov/blog/2016/01/30/computer-science-all.
5. The computer scientist Jeannette Wing is credited with creating and popularizing the term. See her pioneering paper: J. M. Wing (2006). Computational thinking. *Communications of the ACM*, 49(3), 33–35.
6. M. Guzdial (2019, February). Computing education as a foundation for 21st century literacy. In *Proceedings of the 50th ACM Technical Symposium on Computer Science Education*, 502–503. ACM.
7. P. Sengupta, J. S. Kinnebrew, S. Basu, G. Biswas, & D. Clark (2013). Integrating computational thinking with K–12 science education using agent-based computation: A theoretical framework. *Education and Information Technologies*, 18(2), 351–380.
8. M. Heidegger (1954). The question concerning technology. In *Technology and values: Essential readings*, 99–113. Garland.
9. P. Sengupta, A. V. Farris, & M. Wright (2012). From agents to continuous change via aesthetics: Learning mechanics with visual agent-based computational modeling. *Technology, Knowledge and Learning*, 17(1–2), 23–42.
10. Y. B. Kafai, M. L. Franke, C. C. Ching, & J. C. Shih (1998). Game design as an interactive learning environment for fostering students' and teachers' mathematical inquiry. *International Journal of Computers for Mathematical Learning*, 3(2), 149–184.
11. I. E. Harel & S. E. Papert (1991). *Constructionism*. Ablex.
12. S. Papert (1993). *The children's machine: Rethinking school in the age of the computer.* Basic Books.
13. A. V. Farris & P. Sengupta (2016). Democratizing children's computation: Learning computational science as aesthetic experience. *Educational Theory*, 66 (1–2), 279–296.
14. See for example:
S. A. Fincher & A. V. Robins (Eds.). (2019). *The Cambridge handbook of computing education research*. Cambridge University Press.
Khine, M. S. (Ed.). (2018). *Computational thinking in the STEM disciplines: Foundations and research highlights.* Springer.
15. For example, see the following papers:
L. A. Miller (1974). Programming by non-programmers. *International Journal of Man-Machine Studies*, 6(2), 237–260.
A. Robins, J. Rountree, & N. Rountree (2003). Learning and teaching programming: A review and discussion. *Computer Science Education*, 13(2), 137–172.
M. Guzdial, & B. du Boulay (2019). The history of computing education research. *The Cambridge handbook of computing education research*, 11–39.
Please also note that the term "misconception" is deeply problematic: J. P. Smith III, A. A. diSessa, & J. Roschelle (1993). Misconceptions reconceived: A constructivist analysis of knowledge in transition. *Journal of the Learning Sciences*, 115–163.
16. M. Eisenberg, N. Elumeze, M. MacFerrin, & L. Buechley (2009, June). Children's programming, reconsidered: settings, stuff, and surfaces. In *Proceedings of the 8th International Conference on Interaction Design and Children* (1–8). ACM.
17. Y. B. Kafai & Q. Burke (2014). *Connected code: Why children need to learn programming.* MIT Press.
18. See for example: M. G. Ames (2019). *The charisma machine: The life, death, and legacy of One Laptop per Child.* MIT Press.

19. S. Papert (1987). Computer criticism vs. technocentric thinking. *Educational Researcher*, 16(1), 22–30.

20. B. Sherin, A. A. diSessa, & D. Hammer (1993). Dynaturtle revisited: Learning physics through collaborative design of a computer model. *Interactive Learning Environments*, 3(2), 91–118.

21. M. Guzdial (1994). Software-realized scaffolding to facilitate programming for science learning. *Interactive Learning Environments*, 4(1), 1–44.

22. S. Papert (1996). An exploration in the space of mathematics educations. *International Journal of Computers for Mathematical Learning*, 1(1), 95–123.

23. A. A. diSessa (1979). On "learnable" representations of knowledge: A meaning for the computational metaphor. In *Cognitive process instruction*, Franklin Institute Press.

24. M. M. Bakhtin (1981). *The dialogic imagination: Four essays* (Vol. 1). University of Texas Press.

25. D. K. Rosner (2017). *Critical fabulations: Reworking the methods and margins of design*. MIT Press, loc. 2294 of 5141, Kindle.

26. S. Ahmed (2006). *Queer phenomenology: Orientations, objects, others*. Duke University Press.

27. M. G. Ames (2019). *The charisma machine: The life, death, and legacy of One Laptop per Child*, 69. MIT Press.

28. B. Sherin, A. A. diSessa, & D. Hammer (1993). Dynaturtle revisited: Learning physics through collaborative design of a computer model. *Interactive Learning Environments*, 3(2), 91–118.

29. Ames, *The charisma machine*, 2019; see also: S. Vossoughi, P. K. Hooper, & M. Escudé (2016). Making through the lens of culture and power: Toward transformative visions for educational equity. *Harvard Educational Review*, 86(2), 206–232.

30. Vossoughi et al. 2016, Making through the lens, 220.

31. P. Blikstein (2008). Travels in Troy with Freire: Technology as an agent of emancipation. In *Social Justice Education for Teachers*, 205–235. Brill Sense.

32. M. C. Shanahan & R. Bechtel (2019) "We're taking their brilliant minds": Science teacher expertize, meta-discourse, and the challenges of teacher–scientist collaboration. *Science Education*. https://doi.org/10.1002/sce.21550.

33. Papert, Computer criticism vs. technocentric thinking, 1987.

34. D. H. Clements & D. F. Gullo (1984). Effects of computer programming on young children's cognition. *Journal of Educational Psychology*, 76(6), 1051.

35. R. D. Pea & D. M. Kurland (1984). On the cognitive effects of learning computer programming. *New Ideas in Psychology*, 2(2), 137–168.

36. Papert, Computer criticism, 24.

37. Papert, Computer criticism, 1987, 30.

38. See for example: S. Dasgupta, W. Hale, A. Monroy-Hernández, & B. M. Hill (2016, February). Remixing as a pathway to computational thinking. *Proceedings of the 19th ACM Conference on Computer-Supported Cooperative Work & Social Computing*, 1438–1449. ACM.

39. M. M. Bakhtin (1983). *The dialogic imagination: Four essays* (Vol. 1). University of Texas Press.

40. M. Merleau-Ponty (1962). *Phenomenology of perception*, 61. Routledge.

41. L. McMahon (2017). Phenomenology as first-order perception: Speech, vision, and reflection in Merleau-Ponty. In *Perception and its development in Merleau-Ponty's philosophy*, 324. University of Toronto Press.

42. Husserl, Ideas 1, 46. On "radical reflection." E. Husserl (2012). In *Analyses concerning passive and active synthesis: Lectures on transcendental logic* (Vol. 9), Ideas 1, 46. Springer Science & Business Media.

43. McMahon, *Phenomenology as first-order perception*, 2017, 324.

44. J. M. Wing (2017). Computational thinking's influence on research and education for all. *Italian Journal of Educational Technology*, 25(2), 7–14. (p. 8).

45. Merleau-Ponty, *Phenomenology of perception*, 1962.

46. Merleau-Ponty, *Phenomenology of perception*, 1962.

47. S. Grover & R. Pea (2013). Computational thinking in K–12: A review of the state of the field. *Educational Researcher*, 42(1), 38–43.

48. Barber, M. D. (1993). *Guardian of dialogue: Max Scheler's phenomenology, sociology of knowledge, and philosophy of love.* Bucknell University Press.

49. Wing, Computational thinking, 2006, 33.

50. A. V. Aho & J. D. Ullman (1992). *Foundations of computer science.* Computer Science Press.

51. Wing, Computational thinking's influence, 2017, 8.

52. Wing, Computational thinking's influence, 2017, 3718.

53. J. M. Wing (2008). Computational thinking and thinking about computing. *Philosophical Transactions of the Royal Society A: Mathematical, Physical and Engineering Sciences*, 366(1881), 3717–3725, 3718.

54. Wing, Computational thinking, 2008, 3719.

55. J. Locke (1690/1979). *An essay concerning human understanding.* Oxford University Press.

56. Wing, Computational thinking, 2006, 34.

57. D. C. Schmidt (2006). Model-driven engineering. *Computer*, 39(2), 25–31.

58. C. Stewart & K. Shen (2005, May). Performance modeling and system management for multi-component online services. In *Proceedings of the 2nd Conference on Symposium on Networked Systems Design & Implementation*, Vol. 2, 71–84. USENIX Association.

59. Schmidt, Model driven engineering, 2006.

60. J. Wing (2011). Research notebook: Computational thinking—What and why. *The Link Magazine*, 6.

61. Wing, Computational thinking—What and why, 2011.

62. R. Lehrer (2009). Designing to develop disciplinary dispositions. *American Psychologist*, 64(8), 759–771.

63. D. Haraway (1985). Manifesto for cyborgs: Science, technology, and socialist-feminism in the 1980s. *Socialist Review*, 15(2), 65–107.

64. S. G. Harding (1986). *The science question in feminism.* Cornell University Press.

65. Rosner, *Critical fabulations*, 2017, loc. 2281 of 5141, Kindle.

66. Rosner, *Critical fabulations*, 2017, loc. 723, Kindle.

67. S. L. Star (1990). Power, technology and the phenomenology of conventions: On being allergic to onions. *Sociological Review*, 38(1), 26–56, 52.

68. Wing, Computational thinking, 2006, 3717.

69. A. Adam (1998). *Artificial knowing: Gender and the thinking machine.* Routledge.

70. L. M. Terman (1923). *Report of sub-committee of committee on scholarship on student ability.* Stanford University Press.

71. Terman, *Report*, 1923, 28.

72. M. Rose (2005). *The mind at work: Valuing the intelligence of the American worker.* Penguin.

73. J. Streeck (2009). *Gesturecraft.* John Benjamins.

74. J. Margolis (2010). *Stuck in the shallow end: Education, race, and computing*, 156. MIT Press.

75. R. Moses & C. E. Cobb (2002). *Radical equations: Civil rights from Mississippi to the Algebra Project*, 11. Beacon Press.

76. F. A. Pearman, F. C. Curran, B. Fisher, & J. Gardella (2019). Are achievement gaps related to discipline gaps? Evidence from national data. *AERA Open*, 5(4), 1–18. https://doi.org/10.1177/2332858419875440.

77. See for example: S. Y. Lye, & J. H. L. Koh (2014). Review on teaching and learning of computational thinking through programming: What is next for K–12?. *Computers in Human Behavior*, 41, 51–61.

78. J. Derrida (2001). *Writing and difference.* Routledge.

79. D. K. Rosner (2018). *Critical fabulations: Reworking the methods and margins of design.* Design Thinking, Design Theory. MIT Press. loc. 1615, Kindle.

80. S. Turkle & S. Papert (1991). Epistemological pluralism and the revaluation of the concrete. In *Constructionism*, Ablex.

81. J. Jordan, A. Kaplan, J. Miller, I. Silver, & J. Surrey (1991). *Women's growth in connection. Writings from the Stone Center.* Guilford Press. See also: S. Harding (1991). *Whose science? Whose knowledge?: Thinking from women's lives.* Cornell University Press.

82. Papert borrowed this term from Winnicott: D. W. Winnicott (1953). Transitional objects and transitional phenomena—A study of the first not-me possession. *International Journal of Psycho-Analysis,* 34, 89–97.

83. Turkle & Papert, Epistemological pluralism, 1991.

84. U. Wilensky (1991). Abstract meditations on the concrete and concrete implications for mathematics education. In *Constructionism.* Ablex.

85. E. F. Keller (1984). *A feeling for the organism, 10th aniversary edition: The life and work of Barbara McClintock.* Macmillan.

86. M. G. Ames (2018). Hackers, computers, and cooperation: A critical history of Logo and constructionist learning. In: *Proceedings of the ACM on Human-Computer Interaction (CSCW),* Vol. 2, 18.

87. S. Vossoughi, P. K. Hooper, & M. Escudé (2016). Making through the lens of culture and power: Toward transformative visions for educational equity. *Harvard Educational Review,* 86(2), 206–232.

88. Vossoughi, Hooper, & Escudé, Making through the lens, 2016, 220; see also: M. L. Espinoza, & S. Vossoughi (2014). Perceiving learning anew: Social interaction, dignity, and educational rights. *Harvard Educational Review,* 84(3), 285–313.

89. See for example:
T. O'Shea & T. Koschmann (1997). Review, The Children's Machine: Rethinking School in the Age of the Computer (book). *Journal of Learning Sciences,* 6(4), 401–415.
R. Pea & M. Kurland (1984). *Logo programming and the development of planning skills.* Bank Street College of Education, Center for Children and Technology.

90. Two of our previous publications are directly relevant here:
A. C. Dickes & A. V. Farris (2019). Beyond isolated competencies: Computational literacy in an elementary science classroom. In *Critical, transdisciplinary and embodied approaches in STEM education,* 131–149. Springer.
S. Basu, G. Biswas, P. Sengupta, A. Dickes, J. S. Kinnebrew, & D. Clark (2016). Identifying middle school students' challenges in computational thinking-based science learning. *Research and Practice in Technology Enhanced Learning,* 11(1), 13.

91. Shanahan & Bechtel, "We're taking their brilliant minds," 2019.

92. We have illustrated the deep similarities between key facets of computational thinking and scientific inquiry in Sengupta et al. (2013).

93. R. Duschl (2008). Science education in three-part harmony: Balancing conceptual, epistemic, and social learning goals. *Review of Research in Education,* 32(1), 268–291.
See also: A. Pickering (2010). *The mangle of practice: Time, agency, and science.* University of Chicago Press.

94. A. S. Rosebery, M. Ogonowski, M. DiSchino, & B. Warren (2010). "The coat traps all your body heat": Heterogeneity as fundamental to learning. *The Journal of the Learning Sciences,* 19(3), 322–357.

95. E. Manz & E. Suárez (2018). Supporting teachers to negotiate uncertainty for science, students, and teaching. *Science Education,* 102(4), 771–795.

2 A Dialogical Imagination of Coding in STEM

Quests for my own words are in fact quests for a word that is not my own, a word that is more than itself; this is a striving to depart from one's own words, with which nothing essential can be said. I myself can only be a character and not the primary author. The author's quests for his own words are basically quests for genre and style, an authorial position. —M.M. Bakhtin[1]

2.1 Motivation: From Situatedness to Computational Heterogeneity

It is not difficult to trace the intellectual history of constructionism to feminist standpoint scholars who fundamentally questioned the position of knowledge as objective.[2] Rooted in the *situatedness* of knowing that stands in opposition to instrumentalist accounts of learning, constructionist approaches emphasize the construction of public artifacts (that are also personally meaningful to the learner).[3] The immediate epistemological entailment of situatedness, as Wilensky has reminded us, is that the progression of knowing is not from concrete to abstract or from the situated to the removed.[4] Instead, as we understand more deeply, the object of knowing simply becomes progressively more situated in our experiences. The abstract is the unknown, and to become known, it must necessarily become concrete. Higher forms of abstractions, Wilensky posited, mean richer forms of concretion in experience. Bakhtinian lenses of heterogeneity and heteroglossia can help us *see* this process of concretion unfold in ways that can help us avoid the technocentric panopticon.

Another entailment of situatedness is a fundamentally heterogeneous imagination of the learner. Challenging notions of abstractions in technoscience, Haraway argued that if knowledge cannot be separated from contexts—cultural, historical, and personal—then the image of *the knower as an autonomous entity* must also be challenged. Positioning subjectivity at the center of human experiences of knowing and being implies heterogeneous and emergent conceptualizations of human-technology relationships. Haraway's *cyborg* is such an emergent conceptualization in which boundaries between the human and the computer, the mind and the body, and the social and the material worlds are fluid and necessarily transgressed. Ames's critique of the individualistic trajectory of computing and educational computing must also be kept in mind in this regard.[5] This is aligned with a Bakhtinian imagination in which human acts are imagined as "text *in potentia*,"[6] that is, ongoing acts of voicing in which *thingification* and *personification* are inextricably intertwined and the "I" is in constant interrelation with the "other."[7]

We therefore argue for a shift from viewing coding as production of computational artifacts to voicing computational utterances. This shift in metaphors is, in essence, a shift away from an overt reliance on a device-centered discourse of control to expansive imaginations of heterogeneity and heteroglossia. Our concern is that despite epistemological roots in the situatedness of feminist technoscience, constructionist approaches have largely fallen short of challenging technocentrism. Somewhat recursively, our goal here is to offer the Bakhtinian lens as a set of epistemic tools that can more directly counter technocentric imaginations of computing beyond accounts of situatedness. By moving beyond objects and ownership, voicing computational utterances can be seen as an ongoing search for others and otherness. Our proposal implies a *différance*[8]—that is, both difference and deferral—of meaning, rather than foregrounding immediacy and control at the center of experiences of coding. It also indicates a repositioning of the learner from an autonomous entity to a social voice.

In the rest of this chapter, we outline elements of a Bakhtinian framing of coding in K–12 science and STEM contexts that can help us engineer these shifts. We position heterogeneity as the fundamental anchor of this Bakhtinian vision, and identify some key elements: perspectives, addressivity and alterity, and transparency.[9] While these constructs are essential to understanding a Bakhtinian view of language in a general sense, our goal here is to outline them in the context of coding in K–12 science and STEM so that they lay the groundwork for the following chapters, which in turn offer empirical illustrations of these elements of heterogeneity.

2.2   Voicing Code in STEM: A Dialogical Imagination

2.2.1   The Anchor: Voice as Heterogeneity

We begin our journey by centering our attention on the notion of *voice*. Bakhtin argued that *a* voice is an act of coming together, in the forms of hearing, speaking, and co-opting of a multitude of voices.[10] The uniqueness of an utterance is created through ventriloquation, the process by which one or more voices speak through another voice (or a voice type).[11] The voice is thus both porous

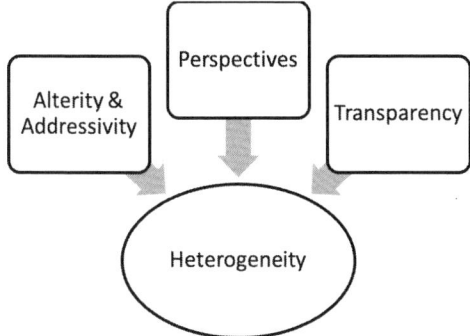

Figure 2.1

A dialogical model of heterogeneity.

and a whole. It is both intentional and historically and socially constrained. Voicing involves a search for words that are not the speaker's: "Quests for my own word are in fact quests for a word that is not my own, a word that is more than myself; this is a striving to depart from one's own words, with which nothing essential can be said."[12]

Understanding what *a* voice represents and how it comes to be, despite its univocal rendition in the form of what gets uttered, thus requires a manifold imagination, akin to Bakhtin's metaphor of refraction.[13] In this image, the meaning of a word emerges through an interplay of the word with the world, in a manner akin to sunlight becoming visible to the human eye through spectral dispersion in the atmosphere. That is, the color of sunlight visible to the eye emerges as the light emanating from the sun as it passes through and optically interacts with various elements that make up the atmosphere, which in turn alters its speed and direction of travel, as well as the colors that eventually are visible to us. So, what we perceive and conceptualize as a "property" or an "attribute" *of* the sun—the color of sunlight—emerges through interactions with myriad elements of the universe beyond the sun. Likewise, the intention of the word gets made and remade as it disperses through the world, even in cases in which such dispersion might be construed to happen within a speaker's mind.

For Bakhtin, a "word" is therefore much more than a symbolic object. It is a *phenomenon*—an unfolding, a "tension-filled environment," where meanings and voices compete and combine with one another. The word is not simply the *container* or *carrier* of meaning, and neither can it be "owned" by anyone. In this view, words are not neutral objects; they always belong to someone else until an *utterance* is created, and when it does become a part of the utterance, it becomes "one's own" only when the speaker populates it with her or his own intention. To quote Bakhtin, it is the *social atmosphere* of the word, refraction through which makes the word meaningful. As Wertsch points out, this is a dynamic imagination of language, in which heterogeneity of and conflict among voices is essential for understanding how univocality emerges.[14] It is also important to note that the essential counterforce to heterogeneity is aesthetics, which Holquist positioned as "the struggle to achieve a whole out of this heterogeneity."[15] This is not merely an additive force, but is rather a synthetic and transformative force, one that lends itself well to Dewey's positioning of aesthetic experience as a fundamental form of human experience. It is aesthetic experience that renders the world meaningful to us,[16] and it can be particularly helpful for advancing our critical phenomenological agenda. We return to this issue in more detail in the following sections.

A fundamental premise of our book can be stated here simply: what is true of the word is also true of code. As with the Bakhtinian utterance, we position computational utterances as elemental pieces of experience that are the sites at which the constancy, historicity, and systematicity enter into contact and struggle with unique, situated performance. This is an emergent imagination of language, in which meaning of an utterance emerges from the interactions between manifold forces, some of which seemingly oppose each other. We define computational utterances as computer models and simulations, embodied and material representations that students and teachers construct to make code meaningful. In this section, we illustrate how the lenses of perspectives, transparency, alterity, and addressivity can help us understand computational utterances.

2.2.2   Dialogical Lenses for Modeling Heterogeneity

2.2.2.1   Perspectives   Our attention to perspectives arises from Bakhtin's positioning of *heteroglossia* and polyphony as essential characteristics that render meaning to any utterance.[17] However, because *heteroglossia* was a term coined by Bakhtin's English translators, its meaning itself is somewhat heteroglossic.[18] In one view, heteroglossia, or *other-languagedness*, refers to the ideologies inherent in the various social languages we partake in our daily lives, such as the language and the inherent ideologies in our professional worlds, age groups, the current decade in time; of our social class, geographical region, family, circle of friends, and so on.[19] More broadly, it can indicate conflicting discourses within any field of linguistic activity (e.g., a national language, a novel, or a specific conversation). Polyphony or *many-voicedness* refers to the collective quality of an individual utterance: by embodying other voices within itself, the utterance creates a dialogic relationship between different voices (the speaker's and the others'). As Bakhtin noted, "Every utterance participates in the 'unitary language' (in its centripetal forces and tendencies) and at the same time partakes of social and historical heteroglossia (the centrifugal, stratifying forces)."[20]

It is not simply vocabularies that distinguish the different social languages that constitute utterances; rather, as Rosebery and colleagues noted quoting Bakhtin,[21] they differ from each other in terms of "specific points of view on the world, forms for conceptualizing the world in words"[22] and as "specific forms for manifesting intentions."[23] Also, the polyphonic layering of voices that renders a novel its rich discursive character can be understood in terms of embedding multiple points of view—of the narrator, the character, and the genre—all within a single voice.[24]

The discursive complexity in STEM classrooms can also be understood in this light, as Rosebery, Warren, and their colleagues at TERC and Boston University have shown through their long-term research project focused on heterogeneity and heteroglossia in the science classroom. Their work shows how the heteroglossic nature of a novel can help us understand the discursive complexity in a science classroom. Here we borrow an example from them that illustrates the importance of points of view in understanding the fundamentally heteroglossic nature of classroom science talk around the word "cold." Heat and cold are common experiences in our daily lives, as well as commonly explored topics in the science classroom.[25] People typically use the word *cold* to refer to a sensory experience. For example, to a 3rd grader holding an ice cube, the ice cube is making her hand cold. Similarly, the word *heat* is typically used in everyday language in connotation to the sensory experience of feeling hot. To a physicist, however, at 32 degrees Fahrenheit an ice cube has a lower temperature than the child's hand; heat energy is thus transferred from their hand to the ice cube. The simple act of holding an ice cube is a complex phenomenon that also involves thinking about the pressure exerted by the hand on the ice, which can lower the melting point of ice and further complicate matters. Furthermore, in physics parlance, both *hot* and *cold* are states of matter that indicate temperature, whereas *heat* is the property of matter (a form of energy) that the temperature represents.

The difference in language use and the implied conceptualizations across these cases can be understood in light of perspectives or points of view that are in play. As Rosebery and colleagues have pointed out, a physicist's perspective offers a fundamentally different way of seeing heat and cold, a fundamentally different point of view. It brings into view a set of interactions and mechanisms of energy transfer, the notion of energy itself, and the notion of pressure. The interpretation of the experience of feeling cold or feeling hot is significantly different, as a result of these perspectival differences, from lay use of the term *cold*. Adopting a psycholinguistic approach,[26] Greeno and van de Sande[27] remind us that physicists' (disciplinary experts') perspectives rely on a set of constraints that serve as conditional relations between situation types. These constraints can be understood as properties and relations of objects and events that constitute a situation from a disciplinary perspective. Thinking about these constraints, in turn, can then offer a disciplinarily authentic way of conceptualizing the situation. Learning then involves developing such perspectival ways of thinking about the relevant phenomena.

There are, of course, several different ways in which we can conceptualize and analyze the role of perspectives in learning that scholars (psychologists, linguists, and sociologists) have long argued for. Rommetveit[28] argued that alignment of perspectives is a necessary condition for understanding the intended meanings of sentences and messages. MacWhinney[29] argued that syntactic features function to signal shifts in perspective, supporting comprehension of sentences. Talmy[30] noted that not only does our spoken language structure our conceptualization of space, but it also structures how we comprehend time. In sociology, several scholars—perhaps most notably, Goffman[31]—have greatly advanced the study of perspectives by illustrating how people's social positioning (*footing*, to use Goffman's term) shapes how they understand themselves in relationship to one another in terms of the social and cultural expectations implicit or explicit in the setting. Ackermann[32] further argues that perspective taking could involve different ways of projecting the *self-in-context*,[33] including taking on different characteristics of others in the situation.

Our point here is not to argue about the utility of some of these ways of thinking about perspectives over others. Instead, we want to emphasize that paying attention to the perspectives in which coding is enmeshed is essential for both understanding and supporting learners' experiences. This is particularly relevant in K–12 STEM classrooms, given the inherent interdisciplinarity of using computing in disciplinary contexts. It can also help us—researchers, designers, and teachers—bridge thinking about *concepts* with thinking about *activities*.[34] As chapter 3 illustrates, doing so can help us see what in a situation is really the "roadblock" for the learner, or conversely, what in the situation is worth paying attention to, from a pedagogical perspective, for bringing about perspectival coherence. After all, it is our *points of view* that decide (both perceptually and conceptually) what we notice in a situation, and understanding, as Greeno and colleagues have argued, typically requires bringing about an alignment between the multiple points of view involved in our interactions with others and the world. It is therefore no surprise that cognitive scientists have argued that both perspective *taking* and perspective *tracking* are central to human sense-making.[35] Following Greeno and van de Sande we henceforth refer to this form of thinking as *perspectival thinking*.[36] The former helps structure the constituents of mental models by linking actions to referents, and the latter helps us link our individual mental models with broader cultural understandings.

Beginning with Papert's Logo, pioneers in computing education have long argued that perspective taking is indeed helpful for learning to code, and we revisit this conversation in more detail in chapter 3. It is, however, appropriate to mention here that the paradigm of computing that we use in this book—*agent-based computing*—is rooted in Papert's vision of the learner being able to take on the perspective of computational agents (e.g., the Logo Turtle). Agent-based computing recasts any phenomenon as an interaction between an agent and elements in its environment (e.g., modeling the motion of an object as an interaction between the object and the surface it is moving along) or as interactions among multiple agents as well as elements in their environments (e.g, modeling ecological interdependence as the dynamic interaction between predators and prey in an ecosystem).[37] In either case, students and teachers interacting with the simulation and the underlying code are prompted to take on the perspectives of different agents or elements in the simulation. This allows them to dive into the phenomenon by taking on a *bottom-up* perspective and experience the phenomenon from the perspective of different agents in the system. At the same time, they are able to take on a *top-down* view by looking at the system-level, aggregate behaviors and outcomes that result from these interactions—for example, graphs of populations of different species in an ecosystem that illustrate their interdependence.

There is now a substantial body of literature that shows that adopting the agent perspective can lead even young learners to develop deep understandings of aggregate-level phenomena in multiple domains such as physics, ecosystems, materials science, chemistry, and so on.[38] While these studies primarily use students' explanations as data, there has been relatively less focus on the microdynamics of perspective taking and perspective tracking in shaping students' explanations, and what this might mean for teachers trying to support the development of these explanations. This is particularly relevant in STEM classrooms, given that multiple perspectives are involved in working with programming languages, disciplinary practices, and spoken languages, especially in collaborative settings. This is (unsurprisingly) a phenomenological agenda:

our goal is to expand and deepen our understanding of the *sense experiences*[39] of students and teachers in engaging with computing in the science classroom. As we begin to see coding as conversations between students (and teachers), the complexities of negotiating multiple perspectives inherent in such conversations offer both opportunities and challenges for engaging with the code and the simulation.[40] Chapter 3 offers an insight into such heterogeneity, and chapters 4 and 5 offer insights into forms of perspective taking that involve negotiating boundaries of the self and the other in ways that are better understood in light of *alterity* and *addressivity*, which we discuss next.

2.2.2.2 Alterity and addressivity   Wertsch[41] presented a phenomenological account of how language is *experienced* from a Bakhtinian perspective. He argued that Bakhtin's dialogic imagination necessitates that we view language not by studying how people "receive" meanings that reside in speakers' utterances, but by focusing on how interlocutors might use texts as tools for thought and create new meanings. Wertsch also pointed out that in Bakhtin's work, central to this image of multivocality is the experience of *alterity* or *otherness*, which can be understood as the dynamic interaction between one voice and another. Bakhtin argued that it is through this ongoing interaction that utterances emerge. Clark and Holquist similarly argued that for Bakhtin, otherness is the ground of all existence and of dialogue, representing "a constant exchange between what is already and what is not yet" (p. 65).[42] To *be*, in Bakhtin's words, is "to be for the other, and through him, for oneself . . . I must find myself in the other, finding the other in me."[43]

One of the foremost scholars of Bakhtin's work, Todorov points out that in order to be understood, every representation of language must put us in contact with its utterer. This is also an experience of otherness or alterity. It is through this process of identifying *who* is speaking that we become *conscious* of language.[44] Therefore both creating an utterance and listening to one can be reconceptualized as negotiations of the boundary between the self and the other, in the same way that Bakhtin problematized the relationship between the "inside" and "outside" of text. For him, boundaries between the origin, the context or referent of text, and its form or structure are porous and fluid, and *language in use* is the heterogeneous act of negotiating these boundaries. Meaning is the coherent shape of these boundaries that emerges through the negotiation, during which elements inside and outside of the text, as well as the speaker and the listener, are in put in conversation with each other. Arguing against viewing text as authoritarian or monologic discourse, Bakhtin noted that "we evaluate our exterior not for ourselves, but for others through others."[45]

The alterity of language makes each voice a consummation of myriad voices and perspectives. Voicing and listening are nested recursively within each other, and to understand an utterance or speech we must learn to listen who are the others speaking through the speaker. An essential element of alterity is therefore the more complex situation of *reported* speech: "of speech within speech, utterance within utterance, and at the same time also speech about speech, utterance about utterance."[46] The simple presence of an addressee does not necessarily make speech dialogical; rather, it is the possibility of commenting or reporting on someone else's speech—the "active reception of other speakers' speech"[47]—that is the essence of dialogicality.

When we look at a line of code as a form of heteroglossic text and at coding as acts of forming computational utterances, many forms of the otherness become explicit and essential for working with code, particularly in the context of K–12 STEM disciplines. For example, addressivity is the inseparable other of an utterance, and pedagogical designs around coding can benefit greatly from enlivening the addressee as an integral part of designing the computational utterance. This becomes evident in chapter 4, in which coding in a math classroom becomes reframed as computational design, and we then see how paying attention to addressivity by involving the user's voice *within* the design process plays an important role in deepening students' engagement with both computational and disciplinary practices. In chapter 6, we present another illustration of alterity, which begins with imagining how a line of code is represented in different forms as part of a computer model in a science classroom. Simply put, a line of code can be represented *within the computational model* as a string of programming commands, comments within the code explaining the meaning of the commands, and visualizations in the form of simulations and graphs. However, teaching and understanding the mathematical relationships represented by the code may necessitate the use of other forms of modeling, such as embodied and physical modeling. Coding, especially in the K–12 science classroom, is not merely the act of creating a computer program that compiles and generates the desired output. It is, instead, a heterogeneous and heteroglossic language that integrates materiality, discourse, and embodied interactions outside the computer with the symbolic world that constitutes the computer model. Code and its *other*—which at the broadest level, is the world outside the computer—are thus deeply intertwined in experience, in the K–12 science classroom, and we will see several images of such alterity in the empirical chapters.

The otherness of code can also play an important role in a critical phenomenological sense. Critical phenomenology points to the ways in which "experience" is not universal; instead, we must learn to recognize how some voices and ways of knowing, feeling, and perceiving are privileged but others tend to be silenced or excluded, particularly in disciplinary and institutional settings.[48] Prominent feminist and post-structuralist critiques of technoscience have identified technological determinism[49] and masculinist notions of a "pure" discipline[50] as deterrents for equity in cultures of computing. At stake here is what should count as code and coding, including possible imaginations of what computational design and computational science can look like in professional practice as well as in the K–12 classroom. A particular critique of technological cultures that we build on in chapter 7, for example, is how relational work—the essential act of caring for, and helping others in complex technology design projects—is institutionally devalued in favor of reifying the myth of individual accomplishment through abstract and reductionist measures of efficacy.[51] Another form of disciplinary expansion is evident in chapter 5, where we consider how computational agents could stand in as *transitional others* and enable preservice teachers to engage in difficult conversations about race and urban segregation in the US. Race talk can become code talk, and relational work can deepen children's engagement with computational work.

2.2.2.3 Ambiguity and transparency   Central to our discussions on perspectives, alterity, and addressivity is the repositioning and reimagining of the human-machine boundary by challenging the orthodoxy of the persistent vision of computing that restricts studies of learning and computer-human interaction to device-level engagements.[52] Haraway's *cyborg* can now be understood as an example of Bakhtin's notion of *hybrid construction*.[53] The boundary between the human and the computer is fluid because, in Bakhtinian parlance, one voice speaks through the other. The metaphorical image here is of intertextuality rather than the literal caricature of a robotic voice ventriloquating through a human. Todorov[54] and Kristeva[55] argued that Bakhtin's "dialogism" and "polyphony" are forms of *intertextuality*, which was defined by Kristeva as follows: "Any text is constructed as a mosaic of quotations; any text is the absorption and transformation of another."[56] If we are to locate the ambiguous boundary between human and computer languages in the intertextuality of code, it follows that we must learn to pay close attention to absorptions and transformations that result when computer languages come in contact with human discourse.

A closer look at the nature of representational work of scientists reveals similar ambiguous boundaries between ideas and the world. Pickering noted that scientific advancement could be understood as a *dance of agency* between the scientist's ideas and the material world that it both acts upon and gets acted upon by. Latour described the work of "designing" scientific representations as a dynamic balance between *representational amplification* and *reduction*. The reducing and amplifying qualities of scientific inscriptions make them sites of conceptual innovation, because they use "the distinctive characteristics of the material world to organize phenomena in ways that spoken language cannot— for example, by collecting records of a range of disparate events onto a single visible surface."[57] The objectivity of a scientific representation relies as much on its *antecedent history*—that is, how it got to be, perhaps as an act of coming together of heterogeneous events and representations as Goodwin stated—as well as, Polanyi argued, its *prospective history*, that is, the conjectures and imaginations of what it might become. In becoming scientifically meaningful, *a* representation emerges through the match and mismatch between multiple representations, that in turn are stable for only a historically bounded period.[58]

Todorov[59] argued that there are three primary forms of discourse. *Literal* discourse signifies without evoking anything (no actual text completely achieves this, according to Bakhtin, despite claims by avant-garde novelists of the "New Novel" movement). In *ambiguous* discourse, several meanings of the same utterance are to be taken on exactly the same level. Syntactic, semantic, and pragmatic ambiguity are all possible. In *transparent* discourse, there is no attention given to the literal meaning (for example, in an allegory). What makes code and computer models particularly amenable to science is that they are both ambiguous and transparent. It is therefore no surprise that computational science involves not only learning to use programming languages in contextually (scientifically) relevant ways but also developing new, interdisciplinary ways of talking about and representing the world. Galison termed scientific simulations "trading zones"—a place where divergent ideas and perspectives are brought together, where theory meets experiments.[60] In the same spirit, Nersessian and her group's long-term cognitive ethnographic research on the creation of scientific knowledge in a biomedical engineering lab has poignantly noted that computational work brings together scientists' and engineers' perspectives. In such settings, dissonance between divergent and different representational traditions must be bridged, which results in the invention of novel representational forms that further scientific knowledge.[61]

Finally, it is important to note that the ambiguity and transparency of computational discourse also lend themselves well to design. Simply put, computational design is the predominant form of activity in the K–12 computational science classroom, as students are typically tasked with designing computer models of scientific phenomena. Herbert Simon's call for the centrality of design in technical professions relied on a model that Schön termed "technical rationality." According to this model, problem solving in engineering and scientific disciplines "is the manipulation of available techniques to achieve chosen ends in the face of manageable constraints."[62] Schön's phenomenological account of scientists and engineers at work—given his emphasis on illustrating their *sense experience* as designers—was strikingly different. He argued that technical problem solving is a *radically incomplete* description of what engineers and scientists do. As scientists and engineers address problems that do not fit known categories, their experience can be better understood as a design process that is artistic in nature and involves engaging in reflective conversations with the situation.[63]

Our view of design is grounded in Schön's phenomenological account of design as a reflective conversation with the situation. Like Schön, we adopt the position that when a designer reflects *in* and *on* their practice, the possible objects of reflection are as varied as the kinds of phenomena at hand and the "systems of knowing-in-practice."[64] The latter includes both the disciplinary lenses and norms that the designer brings to the table. The possibilities of reflection arise in the "action-present"[65]—the zone of time in which action can still impact the situation—and these possibilities are varied in nature. Possible sources of reflection-in-action include, for example, tacit norms underlying a judgment, or the strategies and theories implicit in a behavior. The designer may also reflect on the "feeling for a situation"[66] that has led her or him to construct the particular solution, or may evaluate her or his role within the institutional context. Sometimes, reflection-in-action during design also involves negotiating or shifting between different ways of *seeing as*. Schön argues that engaging in these different modes of reflection is essential for coping with divergent situations in practice.

We believe that Schön's "reflective conversations" usually take the form of a combination of ambiguous and transparent discourse, and there is ample evidence that scientists also engage in such discourse. For example, Ochs and Jacoby's observations showed that physicists' early encounters with new problems often begin with attempts to refine rhetorical elements of the potential explanation of the phenomenon. In the world of science, rhetoric is deeply tied to representational work. For example, what at first is treated as a rhetorical problem—for example, how many dots should be drawn on a graph to be displayed in a conference talk—can evolve into a physics problem—for example, what those dots represent in terms of observed or extrapolated physical processes. In these conversations, Ochs and Jacoby observed that while certain matters of rhetoric remain on a less serious or non-canonical plane, attention to rhetoric is often just a first step in a longer deliberation leading to canonical representations and formulations that later become less ambiguous over time and codified as "physics."

To this end, positioning code and coding as heterogeneous language is an instrumental move on our part. Most immediately, it reveals an essential disciplinary heterogeneity, where coding is at once a language *of* science and a language *about* design. In the first image, coding embodies the "doing" and the "concepts" of science. In Schön's terms, it becomes the *design domain* of scientific work as it combines ways of speaking about and representing the relevant phenomenon from multiple perspectives. In the second image, because coding involves dealing with a programming language that is distinct from the commonly used scientific representations such as equations and graphs, it also serves a *metarepresentational* purpose.[67] Thinking and talking about the meaning of code, as well as creating *other* (e.g., material) representations in order to make the code contextually meaningful, can become a way of reflecting about the *nature* of scientific work and design. Both these dimensions become explicit in our analyses presented in the following chapters.

2.3 Epilogue: In Defense of Heterogeneity

2.3.1 A Critique of Authoritarian Voice

In critiquing monologism and authoritarian voice, Voloshinov / Bakhtin[68] reminded us that:

> History knows no nation whose sacred writings or oral tradition were not to some degree in a language foreign and incomprehensible to the profane. To decipher the mysteries of sacred words was the task meant to be carried out by the priest-philologists. It was on these grounds that ancient philosophy of language was engendered: the Vedic teaching about the word, the Logos of the ancient Greek thinkers, and the biblical philosophy of the word. . . .
> —Bakhtin, 1973, 74.

For Bakhtin, the philosopher and the priest's power comes from their self-declared proximity to the "truth of the word,"[69] a form of authoritarian discourse that is always inaccessible to the rest of society. Bakhtin's dialogical imagination, with a particular emphasis on polyphony and heterogeneity, was fundamentally a challenge to such authoritarian discourse. Critical computing scholars such as Morgan Ames,[70] Safiya Noble,[71] and Lilly Irani[72] have pointed out the dangers of unproblematic adoption of such authoritarian discourse on technology and innovation in terms of exacerbating sociohistorical inequalities. Noble's work unearths how apparently race-neutral algorithms embody and perpetuate racism. Ames reminds us of the dangers of reducing computing education to device-level engagements in the context of challenges with the One Laptop Per Child project. And Irani illustrates the complex interplay of caste, gender, regional identity, and class that underlies practices and experiences of Western and Americo-centric, *colonial*[73] notions of design, technological innovation, and entrepreneurship in the Global South.

Our project here draws inspiration from such critiques of technocentrism and authoritarian discourse on computing. We, however, seek to challenge technocentrism and authoritarian discourse within the microcultures in K–12 STEM classrooms and contexts using an epistemological approach grounded in Bakhtinian heterogeneity. Similar to Bakhtin, the implicit contrast in our work is between an authoritarian image of learning to code and a multivoiced, heteroglossic one. In the former, students and teachers are held captive in their experience of coding as reproduction and recombination of a set of already-known symbolic forms that in turn are understood only by disciplinary experts in computer science. The authoritarian voice here can also be understood as commonly held views of disciplinary authenticity which shape K–12 computing[74] and STEM education.[75] Such reductive views of authenticity primarily rely on a narrow set of experiences and perspectives of disciplinary expertise (e.g., reductive definitions of computational thinking, see Section 1.3) which teachers and students must conform to. In our case, this also reifies a technocentric image in which coding is positioned as device-level engagements, in which the heterogeneity of experience is lost or silenced. For researchers, adhering to such views imply that technological productions (forms of computer code) need to be considered as the primary form of data from which they must infer students' experience. In contrast, we have argued that focusing on perspectives, alterity, and transparency of the experience of code as language—rather than simply looking at code and computational structures and representations themselves—may offer a fundamentally richer imagination of coding. This image centers the experiences and lives of the learners and teachers, and does not frame their experiences as imprints of disciplinary canons. This, in essence, is the shift from computational artifacts to computational utterances.

2.3.2   A Turn Toward Critical Phenomenology

The turn toward utterances is a turn toward experience, and thus, decidedly phenomenological. As Merleau-Ponty argued, a phenomenological agenda relies on *radical reflection* through which we must "rupture our familiarity" with the *sphere of givenness*[76] and the familiar must reveal itself in new ways.[77] Things that once spoke, over time, become buried in our cultural worlds: they may lose their revealing capability over time, hiding essential elements of experience from our view. Our cautionary notes about the unproblematic adoption of "computational abstractions" and "computational thinking" as lenses to look at coding stems from similar concerns. They can subsume and hide the complexity of the experience of coding. It is this form of monologic discourse which Bakhtin argued is an extension of authority. It is important to guard against such discourse, as it might render homogeneous the heterogeneity of our possible experiences of code.

Furthermore, we have argued that our pathway to a *critical* phenomenology of coding is premised on the inherent alterity, ambiguity, and transparency of computing as discourse (for example, see 2.2.2.2). It is an essential reminder that experience is not universal. Especially in the context of technoscience, accounts of experience typically privilege the few with ready access to the inner sanctums of technoscience. Paying attention to heterogeneity in the form of perspectives, alterity, ambiguity, and transparency, can help us question accounts of technocentrism, unsilence critical conversations, and center voices from the margins. This is evident in our work, as we bring to light the following: the complex work of negotiating perspectives even during the earliest steps of modeling, a phenomenon that is usually ignored in technocentric accounts of agent-based modeling (chapter 3); thinking about, interacting with, and designing for an authentic audience—an account that challenges device-centered images of computing (chapter 4); and talking about possible experiences of racialization and inequality in the context of reasoning about simulations of segregation (chapter 5). At the same time, *what* gets voiced through these experiences of alterity and the question of *who* is voicing are also of profound importance. How do teachers with no prior coding experience—whose voices are often ignored in our accounts of science, STEM and computing education—adopt and appropriate coding as an integral part of their science classrooms (chapter 6)? How can racialized students who have been historically left out of disciplinarily rich opportunities for coding find themselves as *authors* of code in STEM classrooms (chapter 7)?

Alongside Bakhtin's arguments, we have also drawn parallels to the scholarship on science studies that present an analogous image of science in practice. Scientific objects—physical or symbolic—are imagined as heterogeneous discourse, their heterogeneity rooted both in their ontogenesis and their yet-unfolded becomings.[78] They are as much carriers of historically grounded meanings as they are tools to imagine new meanings. As Rheinberger noted using Derrida's words, scientific objects represent "a differential typology of forms of iteration"[79] that still seeks elaboration. The objectivity of scientific objects, in this perspective, takes on a *différant* form, because their meanings are both emergent and postponed, unfolding themselves in new (but connected) ways in future discourse.[80] Polanyi further points out that différance is actually rooted in our sense experiences of scientific objects, in that we *trust* scientific objects to have "the independence and power to manifest itself in yet unthought ways in the future."[81] The chapters that follow reveal such *différant* images of code and coding in K–12 STEM contexts, challenging notions of disciplinary homogeneity and inviting us to open the doors of our perception to the myriad becomings of code as heterogeneous utterances in K–12 STEM contexts.

Notes

1. M. M. Bakhtin (1986). *Speech genres and other late essays*, trans. Caryl Emerson and Michael Holquist, 149. University of Texas Press.

2. We have described this history to a certain extent in chapter 1. See section 1.2.

3. E. Ackermann (1996). Perspective-taking and object construction: Two keys to learning. In *Constructionism in practice: designing, thinking, and learning in a digital world*, 25–35. Lawrence Erlbaum.

4. U. Wilensky (1991). Abstract meditations on the concrete and concrete implications for mathematics education. In *Constructionism*. Ablex.

5. M. G. Ames (2019). *The charisma machine: The life, death, and legacy of One Laptop per Child.* MIT Press.

6. M. M. Bakhtin (1976). The problem of text in linguistics, philology and the other human sciences: An essay of philosophical analysis. Earlier publication in *Voprosy Literatury* 10, 281–307. (quoted text is on p. 286). As cited in T. Todorov (1984). *Mikhail Bakhtin: The dialogical principle*, 18, Manchester University Press.

7. Todorov, *Mikhail Bakhtin*, 1984, 18.

8. J. Derrida (2001). *Writing and difference*. Routledge.

9. This is not an exhaustive list. We intend to encourage the field to continue this inquiry.

10. Todorov, *Mikhail Bakhtin*, 1984.

11. J. V. Wertsch (1991). *Voices of the mind: Sociocultural approach to mediated action.* Harvard University Press.

12. Bakhtin, *Speech genres*, 1986, 2307–2309 of 2716, Kindle.

13. M. M. Bakhtin (1981). *The dialogic imagination: Four essays*, Vol. 1. University of Texas Press.

14. Wertsch, Mind as action, 1998.

15. M. M. Bakhtin, M. Holquist, & V. Liapunov (1990). *Art and answerability: Early philosophical essays, xxvi.* University of Texas Press.

16. A. V. Farris & P. Sengupta (2016). Democratizing children's computation: Learning computational science as aesthetic experience. *Educational Theory*, 66(1–2), 279–296.

17. Michael Holquist, editor's glossary, in M.M. Bakhtin (1981). *The dialogic imagination: Four essays*, 428, ed. Michael Holquist, trans. Caryl Emerson and Michael Holquist. University of Texas Press.

18. A. Blackledge & A. Creese (2014). Heteroglossia as practice and pedagogy. In *Heteroglossia as practice and pedagogy*, 1–20, Springer.

19. L. M. Park-Fuller (1986). Voices: Bakhtin's heteroglossia and polyphony, and the performance of narrative literature. *Text and Performance Quarterly*, 7(1), 1–12.

20. Bakhtin, *The dialogic imagination*, 1981, 272.

21. A. S. Rosebery, M. Ogonowski, M. DiSchino, & B. Warren (2010). "The coat traps all your body heat": Heterogeneity as fundamental to learning. *The Journal of the Learning Sciences*, 19(3), 322–357.

22. Bakhtin, *The dialogic imagination*, 1981, 291–292.

23. Bakhtin, *The dialogic imagination*, 1981, 289.

24. Park-Fuller, Voices: Bakhtin's heteroglossia, 1986.

25. Rosebery et al., "The coat traps all your body heat," 2010, 325–326.

26. B. MacWhinney (2005). The emergence of grammar from perspective. In *The grounding of cognition*, ed. D. Pecher & R. Zwann, 198–223. Cambridge University Press.

27. J. G. Greeno & C. van de Sande (2007). Perspectival understanding of conceptions and conceptual growth in interaction. *Educational Psychologist*, 42(1), 9–23. *Note*: We also acknowledge an early conversation with Dr. Rogers Hall for pointing us to this paper.

28. See the following references:

R. Rommetveit (1974). *On message structure*. Wiley.

R. Rommetveit (1987). Meaning, context, and control. *Inquiry*, 30, 77–99.

29. MacWhinney, The emergence of grammar from perspective taking, 2005.

30. L. Talmy (1983). How language structures space. In *Spatial orientation*, 225–282. Springer.

31. E. Goffman (1979). Footing. *Semiotica*, 25(1–2), 1–30.

32. Ackermann, Perspective taking, 1996.

33. G. Hatano & K. Inagaki (1987). Everyday biology and school biology: How do they interact? *Quarterly Newsletter of the Laboratory of Comparative Human Cognition*, 9, 120–128.

34. Greeno & van de Sande, Perspectival understanding, 2007.

35. P. Thagard & K. Verbeurgt (1998). Coherence as constraint satisfaction. *Cognitive Science*, 22 (1), 1–4.

36. Greeno & van de Sande, Perspectival understanding, 2007.

37. U. Wilensky & W. Rand (2015). *An introduction to agent-based modeling: Modeling natural, social, and engineered complex systems with NetLogo*. MIT Press.

38. See http://ccl.northwestern.edu/papers.html for a list of papers.

39. See chapter 1 for a discussion of Merleau-Ponty's notion of sense experience.

40. J. Roschelle & S. D. Teasley (1995). The construction of shared knowledge in collaborative problem solving. In *Computer Supported Collaborative Learning*, 69–97. Springer.

41. Wertsch, *Voices of the mind*, 1991.

42. K. Clark & M. Holquist (1984). *M.M. Bakhtin: Life and works*. Harvard University Press.

43. Bakhtin, quoted by Todorov, p. 96. In T. Todorov, (1984). *Mikhail Bakhtin: The dialogical principle*. Manchester University Press.

44. T. Todorov (1984). *Mikhail Bakhtin: The dialogical principle*. Manchester University Press.

45. M. M. Bakhtin, M. Holquist, & V. Liapunov (1990). *Art and answerability: Early philosophical essays*, p. 33. University of Texas Press.

46. V. N. Voloshinov & M. M. Bakhtin (1986). *Marxism and the philosophy of language*, 115. Harvard University Press.

47. Voloshinov & Bakhtin, *Marxism and the philosophy of language*, 1986, 117.

48. S. Ahmed (2007). A phenomenology of whiteness. *Feminist Theory* 8(2). 149–167. See also: S. Ahmed (2013), *Queer phenomenology*. Duke University Press.

49. D. Haraway (1985). Manifesto for cyborgs: Science, technology, and socialist-feminism in the 1980s. *Socialist Review*, 15(2), 65–107.

50. S. G. Harding (1986). *The science question in feminism*. Cornell University Press.

51. J. K. Fletcher (2001). *Disappearing acts: Gender, power, and relational practice at work*. MIT Press.

52. D. K. Rosner (2017). *Critical fabulations: Reworking the methods and margins of design*. MIT Press, loc. 2281 of 5141, Kindle.

53. Todorov, *Mikhail Bakhtin*, 1984, 73.

54. Todorov, *Mikhail Bakhtin*, 1984.

55. J. Kristeva (1986). Word, dialogue and novel. In *The Kristeva reader*, edited by Toril Moi. Columbia University Press.

56. Kristeva, Word, 1986, 37.

57. C. Goodwin (1994). Professional vision. *American anthropologist*, 96(3), 606–633, (p. 611).

58. Hans-Jörg Rheinberger (2000). Cytoplasmic particles. In *Biographies of scientific objects*, ed. L. Daston. University of Chicago Press.

59. T. Todorov (1982). *Symbolism and interpretation*. Translated by Catherine Porter. Cornell University Press.

60. P. Galison (1996). Computer simulations and the trading zone. In *The disunity of science: Boundaries, contexts, and power*, 118–157. Stanford University Press.

61. S. Chandrasekharan & N. J. Nersessian (2015). Building cognition: The construction of computational representations for scientific discovery. *Cognitive science*, 39(8), 1727–1763.

62. D. A. Schön (1983). *The reflective practitioner: How professionals think in action*, 169. Basic Books.

63. Schön, *The reflective practitioner*, 1983, 170.

64. Schön, *The reflective practitioner*, 1983, 62.

65. Schön, *The reflective practitioner*, 1983, 61.

66. Schön, *The reflective practitioner*, 1983, 62.

67. For an early discussion on the relationship between programming in science classrooms and metarepresentation, see: A. A. DiSessa, & B. L. Sherin (2000). Meta-representation: An introduction, *The Journal of Mathematical Behavior*, 19(4).

68. V. N. Voloshinov (1973). *Marxism and the philosophy of language*. Translated by L. Matejka and I. R. Titunik. Seminar Press.

69. D. Carroll (1983). The alterity of discourse: Form, history, the question of the political in M. M. Bakhtin. *Diacritics* 13(2), 65–83.

70. M. G. Ames (2019). *The charisma machine: The life, death, and legacy of One Laptop per Child*. MIT Press.

71. S. U. Noble (2018). *Algorithms of oppression: How search engines reinforce racism*. NYU Press.

72. L. Irani (2019). *Chasing innovation: Making entrepreneurial citizens in modern India*. Princeton University Press.

73. Irani, *Chasing innovation*, 2019, 217.

74. T. M. Philip, & P. Sengupta (2020). Theories of learning as theories of society: A contrapuntal approach to expanding disciplinary authenticity in computing. *Journal of the Learning Sciences*. Available at: https://doi.org/10.1080/10508406.2020.1828089.

75. M. A. Takeuchi, P. Sengupta, M. C. Shanahan, J. D. Adams, & M. Hachem (2020). Transdisciplinarity in STEM education: A critical review. *Studies in Science Education*, 56(2), 213–253.

76. See our discussion in chapter 1, section 1.3.

77. L. McMahon (2017). Phenomenology as first-order perception: Speech, vision, and reflection in Merleau-Ponty. In *Perception and its development in Merleau-Ponty's phenomenology*, edited by K. Jacobson and J. Russon, 308–337. University of Toronto Press.

78. See Rheinberger, Cytoplasmic particles, 2000, 294.

79. J. Derrida (1993). *Structure, sign, and play in the discourse of the human sciences. A postmodern reader*, 223–242.

80. J. Derrida (2001). *Writing and difference*. Routledge.

81. M. Polanyi (1964). Duke lectures. Quoted in M. G. Greene (1974), *The knower and the known*, 219. University of California Press.

3 Coding and Modeling as Perspectival Work

It's depending on how far from where you are you're from, not how long the roller coaster actually is. —Arnav, 5th grader

3.1 Introduction

Cognition is inherently perspectival.[1] A corollary is that learning to code and thinking computationally must also be perspectival. As we have discussed in chapter 2, this is particularly relevant for agent-based computing, in which much of the coding involves programming properties, behaviors, interactions, and the environment of virtual agents. Not surprisingly, there is a well-established body of research that shows how being able to take the perspective of the virtual agent can be helpful for learning.[2] However, little attention has been paid to the experience of learning to take on the agent perspective, particularly for learners who are beginning their journey as coders. This is the issue we take up in this chapter.

The work of coding and thinking computationally is largely enmeshed in the frame of design. A hallmark of design is its goal-directed nature. However, despite the continuous push toward shaping the work of design toward a particular objective, the experience of design involves an interplay between learning to see more and narrowing the focus on a few key elements relevant to the situation. Design scholars have notably referred to this interplay as the diverge-converge pattern in the *double-diamond design process model*,[3] which we posit can be understood as an interplay between perspectives. As Schön argued, the experience of design involves reimagining the situation by bringing multiple perspectives in contact with one another through reflective conversations.[4] Like the playing of an accordion, many different perspectives bellow in and out, eventually cohering together in a resonant fashion. Educators, psycholinguists, and cognitive scientists have argued that it is the formation of a *coherent* perspective that we typically experience as *an act of understanding*.[5]

Relevant to our work, Nemirovsky and colleagues have argued that modeling in the context of science and mathematics education can be viewed in terms of developing a *tool perspective*.[6] They argue that developing disciplinary expertise involves learning to "see" tools (e.g., material supports for drawing geometric figures) as an implicit aspect of our goal-oriented activities, which involves fusing the perspectives of the symbol and the referent. The tool is not merely a disciplinary artifact; perspectival fusion renders the tool as an inseparable aspect of what counts as a scientific or mathematical representation. This is similar to Pickering's notion of the *mangle* between scientific ideas and the materiality of the physical world, a view in which representational work is inextricably intertwined with conceptual advancement in scientific practice.[7]

What does computational thinking look like in this light? This is the question we investigate in this chapter. In what follows, we embark on a journey with two 5th grade students engaged collaboratively in modeling motion using ViMAP, an agent-based programming and modeling language that we developed specifically for curricular integration in K–12 science classrooms.[8] The episodes we report here illustrate the uncertainties and dilemmas in the interpretive work that the duo engage in, as well as the productive role that a teacher can play in resolving them. Our analysis highlights the importance of viewing computational thinking as perspectival thinking.

3.2   Background: Role of Perspectives in Agent-Based Computing

It has long been argued that thinking perspectivally can be quite helpful in learning to code. Papert famously noted that the Logo Turtle is an *object to think with*.[9] For example, a circle could be understood as a quadratic equation as it is typically represented in high school mathematics. Thinking like a Turtle that traces a circle, however, is quite a different experience. It transforms drawing a circle into an embodied and algorithmic activity: move forward by a small step, then turn a little. And then keep repeating these two "rules" until you come back to where you began. Further reflection reveals that the number of steps depends on the amount of turning: if you are turning by 10 degrees at the end of every step, you will need to take 36 such steps before you come back to the original position, thus completing the circle. Taking the perspective of the Turtle thus leads us to creating the circle as a mathematical object, but the mathematical *experience* is quite different from that of using and interpreting the quadratic equation. More importantly, being able to take on the perspective of the Turtle can enable even elementary students to think mathematically about a circle.

As computing advanced in terms of its hardware infrastructure, we advanced from using a single agent for modeling linear systems, to modeling complex systems and phenomena using *many* (e.g., hundreds or thousands of) agents, or multiagent computational models.[10] Unlike the quadratic equation, which seeks to describe the circle exactly, emergent phenomena are nondeterministic. The *overall* behaviors in such systems arise from the aggregation of many simple and linear interactions between many individual objects or agents, and thus they can be hard to define from the perspective of a single agent only. The result of the aggregation of individual behaviors is often counterintuitive and, in many cases, nonlinear. For example, even when cars generally move forward in a traffic jam, the jam *overall* moves in the backward direction.[11]

A number of studies show that multiagent models can be helpful for understanding emergent phenomena.[12] These studies show that apparent anomalies between the individual behaviors and the overall *emergent* pattern can be resolved conceptually as we develop *multilevel* explanations, that is, explanations that bridge the emergent level with the individual level. To do so, we have to take a deep dive *inside* the phenomenon, and learn to see things and events from the perspectives of both the individual agents and the overall system.

Taking on the perspective of the individual agent, however, is *key* for learning. It is usually the more intuitive place to begin than focusing at the aggregate level. For example, in the case of the traffic jam, the mechanism of how the jam forms becomes clear as we take the perspective of the driver of each car. As I (the driver) see the car ahead of me start moving, I also start my car. But these two events are not simultaneous. By the time my car begins to accelerate and start moving, the car ahead has already accelerated and moved a little farther. This creates a gap, because of the relative acceleration of our cars. This gap then becomes smaller as I begin moving, and eventually catch up to the speed of the car moving ahead of me. But, at the same time, what has just happened to me now has been passed down the traffic lane to the car behind me. That is, this gap that results from the relative acceleration of the car in front compared with that of the car behind keeps getting passed in the backward direction, which then explains why the jam moves backward. Therefore, both views or perspectives are important: the agent-level view that involves thinking carefully about the behavior of a single agent relative to its immediate surroundings, and the aggregate-level perspective, which helps us understand the overall pattern. Noticing the gap between successive cars requires thinking in terms of interactions between at least two or three cars and can be viewed as the local perspective, which in turn relies on taking agent-based perspectives. The overall or global perspective here is a top-down view, which reveals the

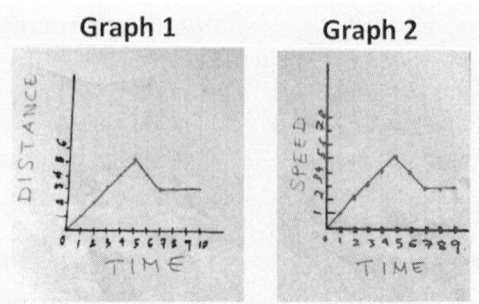

Figure 3.1
Graphs of motion that students were asked to model

direction in which the gap propagates. No one said this better than the late Edith Ackermann: Learning is a dance between local and global perspectives, that is, between "diving in" and "stepping out."[13]

There is now a well-established body of literature on children's agent-based modeling, which shows that adopting the agent perspective can lead novices to develop deep understandings of aggregate-level phenomena in multiple domains such as physics, ecosystems, materials science, chemistry, and so on.[14] But several important questions still remain unanswered for the field. For example, there is little understanding of the process through which learners with no prior background in programming come to successfully adopt the perspective of the agent(s) when they engage with agent-based programming. Another equally important question pertains to the role of teaching. If perspectives are indeed important for learning to code using agent-based modeling, what are some useful ways in which teachers can support the development of perspectival thinking? These are the questions we take on in this chapter.

3.3   The Context

The study took place in a large private university in the mid-southern US as part of a six-day course on computational modeling in science. The course met on the mornings of six consecutive Saturdays during the regular academic year. Twenty students, ages 10–12 (grades 5 and 6), were recruited via a posting on the university website. They were enrolled on a first-come, first-served basis and had no prior programming experience. Pratim and Amy were the instructors for this course.

Our analysis focuses on the discourse between two 10-year-old male students, Arnav and Liam (pseudonyms), as they worked collaboratively in order to program simulations for modeling two cases of motion as represented in two

Figure 3.2
Screenshot of ViMAP (interface) used for the study. This was an early prototype of
ViMAP. Left pane shows the library of command blocks. The middle pane is where the
students create their ViMAP program, and the command block currently being executed
in the simulated output is highlighted in red. The right pane shows the simulated output.
ViMAP uses the NetLogo[15] platform as its simulation engine.

graphs.[16] We present an in-depth analysis of the process of their progressive
conceptualization and symbolization as they engaged in collaborative modeling. We examine the process through which they collaboratively developed a
shared understanding of the task. We show that this process involves bringing
about coherence between multiple perspectives: the object in motion, the computational agent and the visualization produced by the agent, the other student,
and graphs of motion—in order to produce a shared and emergent representation of the problems.

Our analysis reveals how collaborating learners, with prompts from a teacher,
shift across and negotiate multiple perspectives over the course of their interactions, in order to interpret the modeling activity at hand, the relevant programming commands (blocks), and the phenomena to be modeled.

3.4   Unit of Analysis: Perspectives

We present in-depth analysis of 23 minutes of collaboration between the students. The entire activity was video recorded by two cameras focused on the
children and their screens. Additional data were collected as saved files and
screen captures from the students' computers. We created multimodal tran-
scripts of the discourse and coded them turn-by-turn for three points of view
(POVs): the perspective of the agent (the Turtle perspective), the perspective

Table 3.1

Categories and examples of perspectival talk

Perspective	Description	Discourse Example
Turtle	Talk reflects commands for or actions of the computational agent.	"yeah, slow down, forward, and then right"
Object in Motion	Talk is about an imagined object in motion, a roller coaster car.	"it's depending on how far from where you are you're from."
Graph	Talk is about values displayed on the graph or its shape.	"yeah, it tops out at 5"

of an object in motion, and the perspective based on an interpretation of the graph. A description and an example of talk that was coded for each of these perspectives is shown in Table 3.1. Perspectives create an emerging frame through which the students begin to conceptualize motion, and during much of the interaction, the students' interpretations of the two graphs of motion were continually recontextualized as they took on different POVs. In Thagard's terms, these perspectives represent the elements across which coherence is established through a process of shifts and negotiations between these perspectives.[17] Shifts in the dominant perspective indicate important opportunities to learn, and we selected five episodes in which these shifts are occurring. Our analysis focuses both on the students' talk and the teaching moves that both shape and are shaped by these shifts.

The selection of these episodes was guided by the following criteria: a) salience of the shifts in terms of a productive (not necessarily "correct" or "accurate") reconceptualization of the relevant physics, and b) the role of instructional prompts provided by the teacher-researcher. We describe the five episodes of interaction, and within each episode, we present an analysis of a salient, smaller segment of discourse. Additionally, we show a summative analysis of the shifts and coherence in perspectives during the entire interaction.

3.5   Episode 1: Shift to the Agent Perspective

Arnav and Liam began their work together by negotiating with each other how to interpret Graph 1 (Figure 3.1). Arnav wanted to program the ViMAP Turtle to reproduce the shape of the graph, which is a common approach that beginners in physics undertake for interpreting and representing graphs of motion.[18] Liam, however, argued that the graph meant that the distance was gradually going up. The segment of talk that follows is an illustrative example of Episode

1, which is broadly characterized by Arnav's movement to the perspective of the agent, rather than a focus on the shape of the graphs. This particular segment occurred within the second minute of the interaction, immediately after the students had begun their task.

> **Turn 1.1** Arnav: Yeah, but no. Distance 6 (left arm used to approximate slope of first segment of graph).
> **Turn 1.2** Liam: It's GRADUALLY going up.
> **Turn 1.3** Arnav: No, because the next one is speed.
> **Turn 1.4** Liam: No, no, dude. Think about the first graph (points to graph). It's not you-something goes to 6, it's over time we go UP to 6. So then (both?) is what we need to do.

Arnav's interpretation of Graph 1 ignores the role of time over which the change in distance takes place, and he is instead focused on reproducing the overall shape of the first segment of the line in Graph 1. Arnav's interpretation of the graph can also be characterized as an extrinsic graph perspective, because there is no evidence of thinking about the graph as a representation of a process of change of an object. Even though Arnav did consider the actions of the computational agent that would reproduce the shape of the graph, he didn't consider that the line represents a process of change carried out by a particular object in motion. The extrinsic view of the inscription (that is, the graph) was driving the program for the agent.

Liam countered with a different interpretation: the graph meant that "something" is gradually going up to 6. Liam interpreted the graph from a perspective that was intrinsic to the representational features in the graph. That is, he focused on how the shape of the graph came to be, possibly as a representation of an event where something is increasing "over time." He takes on the perspective of actors ("we") to indicate an object whose motion would gradually generate the shape of the graph ("over time we go up to 6"). Also of interest here is that Arnav was unwilling to consider gradual change in distance as a representation of Graph 1, because Graph 2 shows speed, suggesting that he confused the change in speed with the change in distance over time.

Clearly, Arnav and Liam did not share the same initial interpretation of the distance graph, and their disagreement continues nearly until the end of their interactions reported in this chapter. However, despite this disagreement, they begin creating a ViMAP program together. Arnav was not sure whether this was the right move, and yet he agreed to work on the program with Liam. Largely led by Liam, during the next few minutes they used ViMAP command blocks to define individual agent actions that first gradually increased and then decreased the *step size* of the agent. This meant that the Turtle would first accelerate, and then decelerate. Although this is not a "correct" interpretation of what Graph 1 represents in terms of the underlying physics, we believe that specifying individual commands for the Turtle to follow helped the dyad (and especially Arnav) think about the graph as a representation of a process of change. It also created the context for both students to take on the POV of the computational agent, that is, to think about what a computational agent would have to do in order for its actions to represent a case of motion as represented by the graph.

3.6   Episode 2: Teaching as Perspectival Prompts

Arnav and Liam spent the next several minutes creating and editing a program in which the Turtle increased the distance traveled in each step, that is, the agent's step size, then decreased step size, and finally, traveled at the same speed. Pratim, recognizing that many students were confused about what distance meant in Graph 1, called the attention of the entire class to clarify. This occurred at 07:29.

> Pratim: So, this graph here means distance from the starting point. Okay? So this is the distance from the starting point [repeats twice]. So distance going down means you are getting closer to where you started from.

Immediately following Pratim's clarification, Liam began to recognize that there may be some trouble with their existing program. Liam was interpreting the graph as the distance traveled in each step (unit of time); in contrast, the teacher explained that the graph showed how the *total* distance traveled from the starting point was changing with time.

Around minute 10, Liam initiated a conversation with his partner and Amy. He invited Amy for feedback on his work. In her attempts to understand the work, Amy asked Liam and Arnav to explain how their program represented the graph. As evident in Turn 2.7, the students began to suspect that their model might show changes in speed rather than the total distance traveled. Amy prompted the students to move among multiple representational forms: the graph (Turns 2.1, 2.3, 2.18, 2.20), their code (Turn 2.5, 2.10), and the enactment produced by the Turtle (Turn 2.6). The representational forms privileged different explanatory frames: the graph may be interpreted as a description of a process of change or, as Arnav did during Episode 1, an inscription with a particular shape. The code is interpreted as a list of commands specifying specific actions for an agent to follow. The enactment, however, highlights the notion of a process of change because it is a dynamic, changing inscription. Moving *across* the frames promoted comparing frames and resolving discrepancies among them. This interaction occurred during the time 10:31–12:20.

Turn 2.1 Amy: Which graph did you work on first?

Turn 2.2 Arnav: The first one (points to distance-time graph).

Turn 2.3 Amy: The one on the left. So distance is on the *y*-axis. And tell me about what you made.

Turn 2.4 Liam: Um, one speed up forward one speed wait (?).

Turn 2.5 Amy: Can we play your program real slow and you can talk me through it? Okay, just go ahead and slow it down.

Turn 2.6 Arnav: Each time it's supposed to speed up it will eventually speed up to the equivalent of 6.

Turn 2.7 Liam: I think we might have actually been doing speed, I'm not sure.

Turn 2.8 Amy: I think we should think about this some more. Okay, so you're speeding up—

Turn 2.9 Liam: [—and then we slow down].

Turn 2.10 Amy: [Now, how many times have you sped up?] Okay, let's pause for just a second. [pause]

Turn 2.11 Liam: We sped up.

Turn 2.12 Amy: Okay, you sped up.

Turn 2.13 Liam: We sped up 6 times (pointing to sections of code as he talks). Like, we put 5 repeats and then we did it again. We put a right after each one.

Turn 2.14 Amy: Okay.

Turn 2.15 Liam: Uh, and then we had slow down 5 two times, then forward then right and then we just had it going forward right, forward right, like it does.

Turn 2.16 Amy: So at the end, you are neither speeding up nor slowing down. Why is that?

Turn 2.17 Liam: Because it just looks straight.

Turn 2.18 Amy: It just looks straight. So what does a straight horizontal line tell you about speed?

Turn 2.19 Liam: That it's staying at its same speed.

Turn 2.20 Amy: So at this point, which graph do you think you've solved?

Turn 2.21 Arnav (?): Speed.

Note: Arnav and Liam then run the program, and with Amy standing with them, they watch the program run again. The conversation then continues in Episode 3.

In Turn 2.4, Liam initially explained the program simply by listing the commands for the agent. However, Amy attempted to shift the frame from a list of commands to a step-by-step description of what the Turtle was doing in each command by asking them to play the program slowly. That is, she was asking them to slow down the speed of execution of commands using the relevant slider on the ViMAP interface. Upon doing so, the students began to articulate the meaning of the code they produced, in terms of the action of the computational agent, in relation to a graphical representation of the motion.

We can therefore conclude that the role that the instructor played here was to prompt the students' interpretations of different forms of representations of the same phenomenon and to promote coherence across the varying perspectives. *Coherence* across these perspectives emerged through discourse, as Amy specifically prompted them to explain the connections across these representations. We also saw that at the end of the activity, Arnav and Liam began to agree on their interpretation of which graph their ViMAP program was modeling. Amy's request to slow down the execution of the program and talk [her] through it (Turn 2.5) promoted thinking about the individual actions of the Turtle in relation to the graph. This perspective emphasized the enactment of the code in terms of the movement of the Turtle on screen as a process of change. For example, Arnav explains that the Turtle would speed up (that is, accelerate) to 6 units of speed (Turn 2.6). Liam then explains that they sped up 6 times because the command for looping (repeat) specified it so (Turn 2.12). Amy then brings them back to the graph at the end of the conversation (Turn 2.20).

The back and forth between interpretations of the representations and their connections continued throughout the conversation, and so did the turn-taking. Clearly, Arnav and Liam both participated in meaningful ways; their initial differences in interpretation now began to wither away, as Liam began to suspect that they may have been modeling speed (Line 2.7). Coherence across interpretations of representations as well as Arnav and Liam's interpretations now both began to emerge. At the same time, we are able to see how discourse about the design of code as scientific and mathematical representations *amplifies* opportunities for learning through perspectival dances and coherence. Teaching is part of this discourse, which it also shapes through perspectival prompts. Amplification results from both greater conceptual depth and the synthesis of different perspectives. Episode 2 is a deeper dive into the agent perspective compared to Episode 1. Rich in perspectival heterogeneity, it is also an opportunity for synthesis, because it situates multiple perspectives *in relationship* to one another. This episode illustrates the power of perspectival prompts in learning to think computationally about the relationships among code, mathematical representations, and science.

3.7   Episode 3: Teaching as Perspectival Comparisons

Arnav and Liam had now established a joint understanding: their model satisfied Graph 2. They now began work on modeling Graph 1. Our third episode begins with Amy asking Arnav and Liam to delve deeper into their ViMAP code. As we will see, Amy now asks Arnav and Liam to explain how specific regions of their ViMAP Turtle graphics correspond to specific regions of Graph 1. The conversation (16:37–17:53) ensued as follows:

> **Turn 3.1** Amy: Okay, so between . . . so it's getting further away until about time 6. And then what happens between time 6 and time 7?
>
> **Turn 3.2** Arnav: Oh, it starts, it slows down.
>
> **Turn 3.3** Liam: Well, I was counting, and actually, it tops out at 5.
>
> **Turn 3.4** Amy: It tops out at 5, it's really hard to tell isn't it.
>
> **Turn 3.5** Arnav: Yeah, it tops out at 5.
>
> **Turn 3.6** Amy: Okay, so between times 5 and 6, we said, in the graph on the right, it's slowing down. In the graph on the left, "Distance versus Time," is there any difference . . .
>
> **Turn 3.7** Amy: . . . in the distance between times 4 and time 6? If we just compare 4 and 6.
>
> **Turn 3.8** Arnav: No, it's. Yes, there is, but between 3 and 7. Well, yeah, actually, yeah.
>
> **Turn 3.9** Amy: This is a lot to think about, isn't it?
>
> **Turn 3.10** Arnav: Wait. Three and 7 are the same.
>
> **Turn 3.11** Amy: Three and 7 are the same? I am looking at this from a funny angle ((stands)).
>
> **Turn 3.1** Amy: I see it! They ARE the same.
>
> **Turn 3.2** Amy: The same WHAT, though? . . . The same?
>
> **Turn 3.3** Arnav: Distance.
>
> **Turn 3.4** Amy: Distance from?
>
> **Turn 3.5** Arnav: The starting point.
>
> **Turn 3.6** Amy: I think y'all should run with that. Keep talking about it as you work.

The theme of amplification continues in this episode. The representational work of using ViMAP commands to create Turtle graphics became progressively more grounded in perspectival thinking. Across both Episodes 2 and 3, Amy noticed that the students assumed that the two graphs might be equivalent, and her actions were directed toward supporting students to clarify which graph they were modeling. Amy prompted for a deeper dive into the story behind the Distance vs. Time graph. For example, in Turn 3.1 she asked, "It's getting further away until about time 6, and then what happens between time 6 and 7?" Arnav replied that the object that was moving ("it") started and then slowed down (Turn 3.2). This exchange illustrates that there was now a shared understanding of the referent, "it," as an object that moves. Liam then further amplified the response, stating that "*it* tops out at 5."

Arnav's use of the pronoun "it" is worth unpacking. From the perspective of the teacher, it demonstrates how challenging the work of teaching physics using coding can be. We can guess that Liam was using the pronoun "it" either to indicate the height of the graph, or to mean what a physicist would term a *property* of motion (e.g., speed, displacement, etc.). From the perspective of conceptual understanding in physics, since the early 1980s physics education researchers have reported that beginners in physics often find it difficult to differentiate meaningfully between distance, displacement, and speed.[19] We had also encountered such confusions and difficulties in many of our own studies. In order for the coding to proceed any further, the work of teaching here must lead to, or support, conceptual disentanglement. This, in turn, results in more perspectival work, through comparisons across representational forms.

So, in order to clarify this, Amy presented Arnav and Liam with a challenge. In Turn 3.7, she asked them to consider the possibility that two different points on Graph 1 could mean the same thing. In Turn 3.7, Arnav pointed out that "three and 7 are the same." Amy then sought further clarification: "The same WHAT, though?" Arnav then described these two points on the graph as representing the *same distance from the starting point*. Establishing a perspectival anchor, that is, a point of reference for measuring distance, was essential for the students to understand what "distance" meant, as well as what the graph represented. Conceptual amplification was thus attained through perspectival work, because it referentially coordinated aspects of the phenomena that were originally isolated in the students' understanding. This coherence becomes evident in the following and final episode we report in this paper.

3.8   Episode 4: Meaning as Perspectival Coherence

After Amy left, the following exchange occurred between Arnav and Liam (18:19–18:30). In this episode, Arnav extended his new articulation of the meaning of distance from the starting point to an imagined object in motion, acted out in the form of gestures.

> **Turn 4.1** Arnav: It's not depending on how LONG it is ((uses pen as pointer to make an invisible line across the table surface)), it's depending on how far from where you are you're from, not how long the roller coaster actually is.
> **Turn 4.2** Liam: Like he said, it's how far FROM (interrupted by Arnav)
> **Turn 4.3** Arnav: No, so, if you're here ((right hand in front of chin))… and then you do a loop ((half-circle upward motion)), and you come back ((half-circle downward motion)), you'll be pretty much at the same distance from the starting point.

The ViMAP model that Arnav and Liam produced represented the changes in displacement over time, which evidences an agent perspective of getting farther from, then closer to, a point in space. In Turn 4.3, Arnav used the example of a loop, not as in the form of a computational abstraction, but similarly to a vertical circular segment of a roller coaster ride, to explain what was going on in the model. His explanation, reenacted by Amy, is shown in Figure 3.3. His embodied definition and gestural enactment demonstrate the shift in his reasoning about the motion from the perspective of the object in motion. The object is at the same position in the beginning (time = 3) and end (time = 7) of the loop. This merging of symbolic forms with an embodied action is not merely a perspectival fusion; it is also an important conceptual achievement. Arnav described one possible motion—a circular motion—*from* the perspective of the object in motion, that could "fit" the given graph. The object perspective grounded the generality of the graph in a specific case of motion, which in turn could be represented by the agent-based model. In the words of Nemirovsky and colleagues, Arnav thus creates a *fusion*[20] that merges his understanding of the graph and the ViMAP program with his actions in gestural space.

Figure 3.3
Reenactment of Arnav's gestures during his explanation in Turn 4.3

The dance between generality and specificity, which is at the heart of the practice of modeling in science,[21] is also evident in Arnav's statement. Specifically, his construction fused inanimate physics entities with flexibly construed animate objects. Throughout this segment, Arnav's egocentric perspective was merged with the object perspective. Arnav and the (imagined) object in motion were conjoined in simultaneous, multiple constructed worlds: the here and now of the interaction, the visual representation, and the imagined physical processes. In this utterance, the "physics" entities are distances and points, articulated by "how long it is" (where "it" refers to the length of the roller coaster track), and "how far from where you are you're from," in which "where you're from" is a point, and "how far you are from where you're from" is a distance. Grammatically, the verbs used in the utterance ("is" and "are") are in simple present tense. Semantically, as pointed out by Ochs and colleagues in their analysis of physicists' talk,[22] these verb forms allow the embodied action to transcend time and the local setting.

This was a pivotal segment in Arnav and Liam's journey. The conceptual dilemmas associated with what the agent stood for, how it represented the graph, and what the graph meant in terms of a hypothetical real-world scenario of motion were now resolved through the perspectival coherence evident in this segment. Arnav and Liam then proceeded with their modeling work, deftly identifying how ViMAP commands represented specific segments of the graphs and keeping in mind the appropriate physics. They were now able to move fluidly between the different perspectives, and were using these perspectival explanations to justify their choice of code to one another. For example, the following conversation occurs in minute 19, as the duo decide to further refine their model of Graph 1 (the Distance vs. Time graph):

> **Turn 5.1** Arnav: So we have to put it for 6 seconds ((adds repeating loop with a parameter of 6 seconds))...forward 1, speed up 1 ((adds a forward command in the loop)).
> **Turn 5.2** Liam: Forward 1, speed up 1?
> **Turn 5.3** Arnav: Yeah, because... Actually, we don't need the speed up, because, see, each point on that [distance] graph is one second, each point is one second, and all of them are the same length—we don't need a speed up.

Liam and Arnav referred back to the graphs, and then discussed the appropriateness of relevant ViMAP commands, "Forward (Step-size)" and "Speed-up (Change-in-Step-size)," for modeling the distance graph. In doing so, they coordinated the change in distance as represented on the *y*-axis and time (*x*-axis) with the motion of an object. Arnav noticed his mistake and clarified that their ViMAP distance model did not require acceleration because the length of the line (in the graph) during each second was the same, thereby implying that the object traveled the same distance during each second. This reflects coherence between a graph perspective ("each point on that graph" and "all of them are the same length"), an agent-perspective ("we don't need a speed up"), and an implied object perspective. For the first time in the activity, Arnav computationally parsed the difference between moving forward at the same speed and speeding up. He understood that in order to model the total distance traveled by an object, he did not need to vary the distance traveled in each step. Instead, getting farther from the initial position could be represented by the aggregation of several identical steps—"all of them are the same length" (Turn 5.3). Their model showed the agent getting farther, as discussed, then getting closer to the initial position by taking identical steps in the opposite direction, and staying the same distance from the initial position by not moving at all.

3.9   Reflections

3.9.1   Coding Science as Perspectival Thinking

This chapter shows that when students engage in collaborative, agent-based programming in order to model a scientific phenomenon, the development of computational thinking and learning physics co-occur through negotiations of multiple perspectives or POVs. Coding science involves negotiating multiple representational systems: a programming language, canonical representations grounded in disciplinary practices (e.g., the graphs of motion), and the hypothetical or real-world situation that is being modeled. Moving among these representations and connecting across them promoted shifts and negotiations in the points of view, especially as the two collaborators clarified their own interpretations to each other, and to Amy. A timeline of these perspectival shifts is shown in Figure 3.4.

Our analysis also illustrates how the role of teaching can be construed through the lens of perspectival thinking. Thinking carefully about perspectives and perspectival prompts can be particularly helpful for teachers in the context of supporting students' computational modeling and coding in a science classroom. The work of teaching, as evident here, can be understood as appropriately prompting for perspectival shifts. Arnav and Liam's fluency with interpreting the graph depended not only on having a clearer understanding of what distance meant, but also on actually imagining an object in motion from the perspective of the object, and what it would mean to get farther away and closer to the initial position. This is the *coherent* state that emerges toward the end of the episodes reported here (minutes 18–19), that is, a state of explanations in which these perspectives are all fused with one another. The role that teaching played was to facilitate shifts between these perspectives.

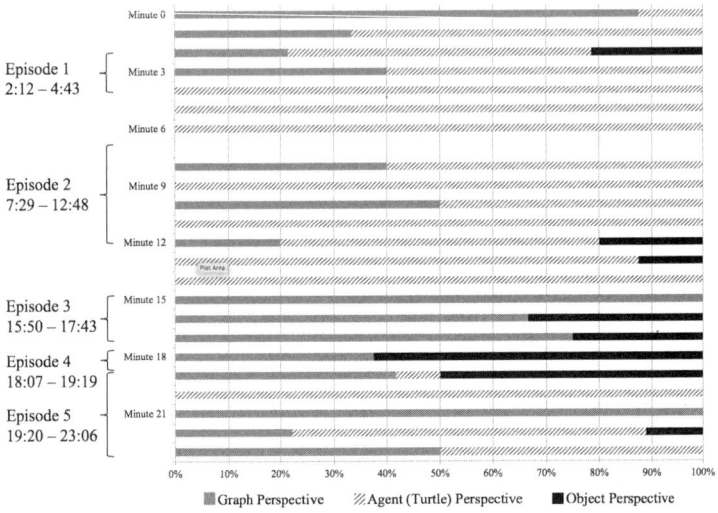

Figure 3.4

A chronology of perspectives

Similar to previous studies using agent-based modeling,[23] our study shows that the agent perspective can indeed play a productive role in understanding the relevant scientific concepts; however, we also show that this perspective needs to be negotiated with others for conceptual growth. These other perspectives included the children's egocentric perspectives, perspectives based on the graphs, and the perspective of the (imagined) physical entity in motion. Achieving coherence between all these perspectives enabled the learners to bridge what is happening now (that is, the instantaneous position and speed of the object in motion) with what has happened until now (that is, previous changes in the object's position and speed). This in turn enables them to begin to describe motion as a process of continuous change—a feat that is challenging for even college-level physics learners.[24]

3.9.2   Theory into Action: Implications for Design

This study had a profound impact on the design of the ViMAP programming language as a modeling environment, as well as on the design of curricular activities around ViMAP for the science classroom. It is representative of our experience with K–12 students (elementary through high school) who often struggle with similar issues while modeling motion using ViMAP or other Logo derivative languages. It is indeed true that they can interpret and under-

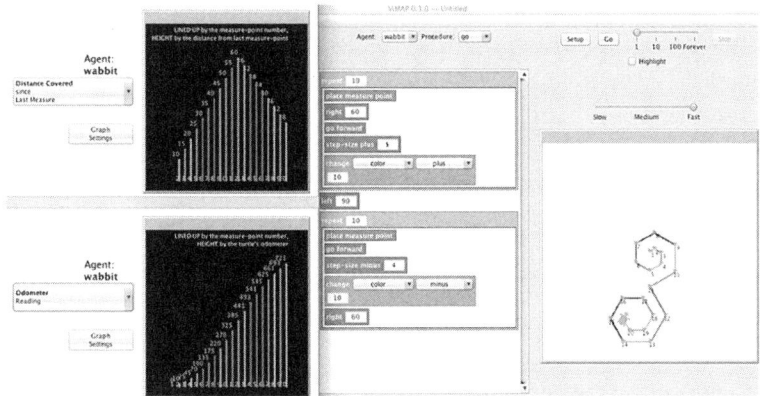

Figure 3.5
Screenshot of current ViMAP interface. The superposed window on the left shows the graphing interface, and the window on the right shows the ViMAP code and the simulated Turtle graphics. Each bar in the graphs is color-matched with the corresponding step of the Turtle in the simulation.

stand the programming commands without extensive difficulties at a semantic level, but the journey from interpreting a *step size* as the size of a Turtle's step, to using *step size* as a variable to represent motion as a process of continuous change, is quite an arduous one.

This study gave us an insight into an important type of challenge experienced by the students: perspectival thinking. It also alerted us to what would soon become an indispensable form of programming and modeling activity within the ViMAP environment: graphing and measurement. A screenshot of the current version of the ViMAP interface is shown in Figure 3.5. ViMAP programming commands include commands for initiating and repeating *measures*, as well as functionalities that allow students to select a variable to be measured.

Programming the behavior of agents requires that learners define an event computationally, by specifying the actions of and the interactions between the agent(s) that will constitute the event. The state of the simulation, at any instant, represents a single event in the form of spatialized representations of agent actions and interactions. To *run* the simulation, these events are repeated by the number of times specified by the user. The graphs present a complementary view by displaying how relevant variables in the simulation change over time. By engaging in iterative cycles of building, sharing, refining, and verifying ViMAP models, students refine their understanding of what actions and interactions of agents optimally constitute a *unit* event. These events are iterated as the simulation runs, thus generating the graph, which in turn provides an aggregate (bird's eye) view of the events. This allows learners to negotiate the agent perspective with other complementary perspectives, as we have illustrated in this chapter.

On a more pragmatic level, all K–12 teachers with whom we have worked in the US and Canada have reported that their students experience difficulties in interpreting and constructing mathematical measures (e.g., units of measurement and graphs) in their regular science classes.[25] Particularly in the US, this has a special significance, because graphing and measurement figure prominently in standardized assessments. A technocentric framing here would be that ViMAP is a tool that can help students overcome these challenges because of the perspectival complementarity of graphs and ViMAP code. A more phenomenological account, on the other hand, would entail paying careful attention to *experiences* of teaching and learning, of which perspectival thinking is an important element. Our phenomenological orientation also implies that we pay attention to the design of learning activities from the teachers' perspectives as they attempt to integrate modeling, coding, and measurement in the science classroom. This is no simple act; besides the degree of familiarity that teachers may already have with coding, the complexity of coding science in the classroom also involves thinking carefully about *materiality* beyond the digital world. We take up these issues in chapters 4 and 6.

3.10   Epilogue: Role of Perspectives beyond the Agent-Based Paradigm

It might be tempting to assume that the notion of perspectival thinking is emblematic and useful only in the context of agent-based computing. Perspectives or points of view play a fundamental role in our *experiences* of computing, regardless of the genre of computing. Here we present a simple but powerful example of how points of view play essential roles in developing understandings of fundamental elements of computer programming, such as developing *data abstractions*. Data abstraction is a fundamental practice in computing because it is essential for understanding and representing compound data by combining simple data objects.[26] It is also deeply synergistic with scientific modeling.[27] Thinking about data abstractions and learning to abstract data computationally involves constructing computational objects that have several parts. This is as much analogical as it is logical, because the components of a computational object correspond to the different aspects that constitute the real-world phenomenon being represented via the compound data.

Ableson, Sussman, and Sussman have illustrated how constructing a data abstraction requires structuring our programs so that they use data in such a way "as to make no assumptions about the data that are not strictly necessary for performing the task at hand."[28] This, in turn, involves taking the perspectives of both the lower-level data representations that the programming language relies on, and the real-world context that the compound object is being designed to represent. For the example of representing a rational number, Ableson and colleagues have noted that creating a compound data object—a *pair*—involves the analogy of gluing together a numerator and denominator. The computer program can then manipulate the *pair* in a way that would be consistent with regarding a rational number as a single conceptual unit.

What makes a pair a rational number, however, is also contingent upon satisfying the following condition: "We need to guarantee that, if we construct a rational number x from a pair of integers n and d, then computationally extracting the numerator (n) and the denominator (d) of x and dividing them should yield the same result as dividing n by d."[29] In perspectival terms, Greeno and van de Sande would argue that the relationship between n and d serves as a constraint—a conditional relationship between situation types—that must be satisfied for the pair to be mathematically meaningful. Ableson and colleagues further noted that the notion of a pair also relies on a more general data structure, a *list*, which enables the use of more general computational operations such as reversing the order of the elements in a list, combining different elements in a list, and so on, to represent a pair.[30] Here again, from a perspectival stance, these structural affordances of a list create further constraints that must be satisfied alongside the relationship between n and d, in order for the data object to be meaningful in computer science. And finally, it is also important to note that understanding how to represent a pair involves another set of constraints in terms of understanding of how programming commands that are specific to the programming language being used (e.g., LISP) and how they access different hardware locations in the computer (e.g., registers).[31]

Our goal in redescribing Ableson and colleagues' classic textualization of how to create and understand data abstractions is to identify the different *points of view* that may be involved in the construction of even a single computational entity. These points of view rest in different disciplinary spaces (e.g., mathematics and computer science), and creating the data abstraction of *a pair* thus involves bringing about coherence between them. In Greeno and van de Sande's terms, this in turn would involve satisfying the constraints that we identified in the previous paragraph. In another sense, as Ableson and colleagues also point out, points of view also involve being able to conceptualize computational representations as both *objects* and *manipulations*—that is, objects that can be manipulated in a manner that is analogous to concrete operations in the real world (e.g., gluing together).

The overall image that emerges here is that of the heterogeneity of perspectives, and complementarily, the heterogeneous nature of perspectival coherence. This is deeply resonant with the story of Arnav and Liam. In an even more general sense, following Nemirovsky and colleagues, we had argued earlier that learning to code and learning to use code as a way to model science can be viewed in terms of developing a *tool perspective*.[32] This includes viewing the tool as an implicit aspect of our goal-oriented activities, often fusing the perspectives of (and blurring the distinction between) the symbol and the referent, as well as viewing the tool as an object of reflection. This is analogous to conceptualizations of layers of computational abstractions (see section 1.3.1), some of which remain implicit whereas others become explicit in the work of *designing* data abstractions. The experience of taking on and developing the tool perspective is less well defined and thus needs to be studied more carefully in computing education, particularly in contexts of K–12 science. Designing a compound data object for modeling scientific phenomena involves emulating the sensitivity of the object to certain aspects of the phenomena relative to others, ascertaining conditions under which it is useful, and so on. Given the constraint-driven nature of the tasks, as we have explained in this section, we hope that computing education researchers will pay attention to the role of perspectival thinking in future work.

To summarize: Our overall point here is that paying attention to perspectival thinking is essential for untethering us from the image of *device-level engagement* that is at the center of our overall critique of educational computing.[33] This includes paying attention to both the heterogeneity of and coherence between the manifold *points of view* inherent in computational science, and it is essential for revealing the complex and ill-defined nature of the experience of computational abstractions. Furthermore, from a pedagogical perspective, paying attention to perspectival thinking can actually help alleviate students' conceptual difficulties in learning to interpret code and computational science.

Notes

1. B. MacWhinney (2005). The emergence of grammar from perspective. In *The grounding of cognition*, ed. D. Pecher & R. Zwann, 198–223. Cambridge University Press. See also: J. G. Greeno & C. van de Sande (2007). Perspectival understanding of conceptions and conceptual growth in interaction. *Educational Psychologist*, 42(1), 9–23.
2. See sections 2.2 and 2.3 for references.
3. This model was proposed by the British Design Council in 2005: Design Council (2005). The "double-diamond" design process model. http://www.designcouncil.org.uk/designprocess. See also: D. A. Norman (2013). *The design of everyday things*, 220. Basic Books.

4. D. A. Schön (1984). *The reflective practitioner: How professionals think in action*, Vol. 5126. Basic Books.

5. C. C. van de Sande & J. G. Greeno (2012). Achieving alignment of perspectival framings in problem-solving discourse. *The Journal of the Learning Sciences*, 21(1), 1–44. See also: J. G. Greeno & B. MacWhinney (2006). Learning as perspective taking: Conceptual alignment in the classroom. In *ICLS 2006: Proceedings of the 7th International Conference of the Learning Sciences*, 930–931. International Society of the Learning Sciences.

6. See: R. Nemirovsky, C. Tierney, & T. Wright (1998). Body motion and graphing. *Cognition and Instruction*, 16(2), 119–172.

7. A. Pickering (1995). *The mangle of practice: Time, agency, and science*. University of Chicago Press.

8. P. Sengupta, A. Dickes, A. Farris, A. Karan, A., D. Martin, & M. Wright (2015). Education: Programming in K–12 science classrooms. *Communications of the ACM*, 58(11), 33–35.

9. S. Papert (1980). *Mindstorms: Children, computers, and powerful ideas*, 11. Basic Books.

10. S. Tisue & U. Wilensky (2004). NetLogo: A simple environment for modeling complexity. In *Proceedings of the International Conference on Complex Systems*, Vol. 21, 16–21.

11. U. Wilensky & M. Resnick (1999). Thinking in levels: A dynamic systems approach to making sense of the world. *Journal of Science Education and Technology*, 8(1), 3–19.

12. For a review, please see: A. C. Dickes, P. Sengupta, A. V. Farris, & S. Basu (2016). Development of mechanistic reasoning and multi-level explanations of ecology in 3rd grade using agent-based models. *Science Education*, 100(4), 734–776.

13. E. Ackermann (1996). Perspective-taking and object construction: Two keys to learning. In *Constructionism in practice: Designing, thinking, and learning in a digital world*. Lawrence Erlbaum, 25–35.

14. Please see http://ccl.northwestern.edu/papers.shtml for a list of papers.

15. U. Wilensky (1999). NetLogo. http://ccl.northwestern.edu/netlogo/. Center for Connected Learning and Computer-Based Modeling, Northwestern University.

16. Our analysis builds on an earlier short paper published in ICLS 2014, in which we presented excerpts from the data and a shorter analysis of perspectival thinking. See: A. V. Farris & P. Sengupta (2014, January). Perspectival computational thinking for learning physics: A case study of collaborative agent-based modeling. In *ICLS 2014: Proceedings of the International Conference of the Learning Sciences*, Vol. 2, 1102–1106. International Society of the Learning Sciences.

17. P. Thagard & K. Verbeurgt (1998). Coherence as constraint satisfaction. *Cognitive Science*, 22 (1), 1–24.

18. L. C. McDermott, M. L. Rosenquist, & E. H. Van Zee (1987). Student difficulties in connecting graphs and physics: Examples from kinematics. *American Journal of Physics*, 55(6), 503–513.

19. D. T. L. McDermott & D. Trowbridge (1981). Investigation of student understanding of the concept of acceleration in one dimension. *American Journal of Physics*, 49(3), 242–253.

20. Nemirovsky, Tierney, & Wright. Body motion, 1998.

21. E. Ochs, P. Gonzales, & S. Jacoby (1996). "When I come down I'm in the domain state": Grammar and graphic representation in the interpretive activity of physicists. *Studies in Interactional Sociolinguistics*, 13, 328–369.

22. Ochs, Gonzales, & Jacoby, "When I come down I'm in the domain state," 1996.

23. U. Wilensky & K. Reisman (2006). Thinking like a wolf, a sheep, or a firefly: Learning biology through constructing and testing computational theories—An embodied modeling approach. *Cognition and Instruction*, 24(2), 171–209.

24. D. I. Dykstra Jr & D. R. Sweet (2009). Conceptual development about motion and force in elementary and middle school students. *American Journal of Physics*, 77(5), 468–476. See also: Note 4.

25. See also: P. Sengupta, B. Brown, K. Rushton, & M. C. Shanahan (2018). Reframing coding as "mathematization" in the K–12 classroom: Views from teacher professional learning. *Alberta Science Education Journal*, 45(2), 28–36.

26. H. Ableson, G. J. Sussman, & J. Sussman (1996). *Structure and interpretation of computer programs*. MIT Press.

27. P. Sengupta, J. S. Kinnebrew, S. Basu, G. Biswas, & D. Clark (2013). Integrating computational thinking with K–12 science education using agent-based computation: A theoretical framework. *Education and Information Technologies*, 18(2), 351–380.

28. Ableson, Sussman & Sussman, *Structure and interpretation*, 1996, 83.

29. Ableson, Sussman & Sussman, *Structure and interpretation*, 1996, 122.

30. Ableson, Sussman & Sussman, *Structure and interpretation*, 1996, 102–103.

31. Ableson, Sussman & Sussman, *Structure and interpretation*, 1996, 122.

32. Nemirovsky, Tierney, & Wright, Body motion, 1998.

33. See section 1.2.

4 Addressivity in Computational Design

4.1 Beyond Individualist Notions of Competence

Historically, issues of children's power and agency were at the center of the concerns of early pioneers of educational computing. Rather than the computer programming the child, Papert noted, the child must be positioned as the creator.[1] DiSessa's notion of *regime of competence*, which relies on the learner establishing a committed relationship with an activity system, further extends this view. The notion of a committed relationship in turn entails "a feeling of ownership, personal connection, and competence such that extended engagement in those activities is perceived to be a natural extension of ourselves."[2]

Critical technology scholars have, however, raised concerns about the history of computing education being rooted in individualism. As Ames argued, "the social imaginary of the naturally creative yearner frames individualized creativity as innate."[3] In this chapter, we offer a different account. We illustrate how looking through a Bakhtinian lens can bring into focus how competence in computational design is dialogically established and how specific forms of dialogue with *others* can play important roles in engaging learners in designing computational artifacts. It can take us beyond the myopia of device-level engagement and individualist notions that largely undergird educational computing.[4]

Drawing upon Bakhtin's notion of addressivity, we position computational design as fundamentally intersubjective. In Bakhtin's heteroglossic framing of language, addressivity is the constant state of being addressed and being in the process of answering. By positioning addressivity at the center of computational design, we also argue for shifting away from the computational artifact as the centerpiece of constructionist education to a more phenomenological focus on computational utterances. Fundamental to this shift is the role of *otherness* in the form of involving the *relevant other*—that is, the authentic user—*within* the design experience, as our empirical journey in a 4th grade mathematics classroom illustrates next.

4.2   From Artifacts to Utterances: Intersubjectivity, Addressivity, and Publicness

Phenomenologists such as Heidegger positioned Being as essentially intersubjective, noting that *Being* is essentially *Being-with*.[5] That is, the substantiveness of our experience cannot be understood in absence of ongoing involvements of our Being, that is, relationships our Being is *always* a part of—"the "in-order-to," the "for-the-sake-of," and the "with-which" of an involvement."[6] In Bakhtin's work, intersubjectivity is strongly reflected in his emphasis on intertextuality and on the fundamentally multivoiced nature of speech. The heteroglossic image of intertextuality that we introduced in chapter 2 represents refraction, in which the meaning of a word emerges as it is refracted through other words, essential to which is the notion of alterity or otherness. Being in the world *with others* involves ongoing *addressivity*, that is, being "responsible" or "answerable" toward others.[7]

In the context of educational computing, scholars who study computer supported collaborative learning (CSCL) have paid attention to intersubjectivity. Suthers[8] argued that intersubjective meaning-making takes place when multiple participants contribute to the creation or design of an artifact that brings together distinct but interrelated interpretations. Here, intersubjectivity is positioned as a participatory process within which (different) views are enacted by each individual without necessarily being mutually accepted. This is different from what we described in chapter 3, in which collaborative modeling and programming led to the establishment of a common ground between two students. Simply put, collaboration may lead to an establishment of a common ground, whereas intersubjectivity might not. Collaboration brings to light contexts in which interpretive differences, rather than agreement, may become meaningful and essential for joint work to proceed.

But from an epistemological standpoint, we stand to gain something else, too. Publicness is an aspect of the experience of learning that has not yet received adequate attention in the constructionist literature, despite Papert's early emphasis on the creation of a public artifact as central to learning rooted in the example of Brazilian samba schools and the culminating carnival.[9] In Papertian parlance, the notion of a public artifact is a device-centered approximation of this carnivalesque imagination, typically positioned as the culminating element of constructionist learning experiences. In contrast, foregrounding the fundamentally heterogeneous and multivoiced nature of computational design offers us an opportunity to examine publicness beyond the emphasis on creating computational artifacts. It brings into focus the voices and interactions with others involved in computational design. In doing so, it makes visible the work of design by making both *use* and the *users* visible. This is an image of Bakhtinian addressivity, as it makes the user visible and present *during* the design experience, rather than making the interaction with users a one-time event.

In countering technocentrism, it is therefore not sufficient to simply say that coding is dialogical without adequately emphasizing its multivoicedness. Intersubjectivity highlights working with others, and addressivity brings to our attention the importance of ongoing interactions with authentic users (and teachers, who also sometimes play the role of users). Publicness is not in the artifact, but is rather in the voicing of a *computational utterance*. In this image, the computational artifact is part of an ongoing conversation beyond the device-level engagement rather than being an end in itself. As we show in our analysis, what becomes public then is not merely the artifact but is rather the disciplinary practices at the heart of computational design that shaped the artifact. The vision of computational design that we present here expands the scope of coding beyond dragging and dropping programming blocks only to be seen by a pair of silent eyes. In our study, for code to become usable and understood, it needs material support (e.g., mechanical devices and user guides); it is also *heard* in conversations, discussed, debated, and gesturally enacted. The learners are also working with *the relevant others* that include members of the concerned public (the users), group members, classmates, and the classroom teacher. Meaning is negotiated between designers (students), the teacher, and the intended users, and assembled by bringing together software and the physical world through mathematical reasoning and representations. Last but not least, interactions with authentic users—that is, users from the "real world" with a legitimate claim for using the designed artifact—become just as much a part of code and coding as the student-designers' competence with the programming language. This chapter shares the story of how a 4th grade math classroom became such a multivoiced space.

4.3 Being in the World and Designing for Others

4.3.1 Intersubjectivity and Collaboration

In the capstone learning activity reported here, 4th grade students worked in dyads to design a mathematical machine for generating geometric shapes. Each machine consisted of a virtual and a physical component. The virtual component was a ViMAP program, which learners constructed using block programming commands. The physical component involved two independent simple machines (e.g., pulleys). Each member of the dyad designed one of these machines to control the reading on a distance (infrared) sensor. Each sensor in turn controlled either the movement or rotation of the ViMAP Turtle.

Intersubjectivity and collaboration are intertwined in this activity. As Suthers noted, intersubjective meaning-making can be supported when learning is distributed across individuals and information artifacts through and with which they interact, which in turn creates opportunuties for each member of a group to bring in their individual voices and perspectives. In our case, in designing the simple machines for controlling the readings on one of the sensors, each learner experiences the opportunity to enact their own ideas independently. Intersubjectivity, then, focuses our attention to the *different* contributions of each member, and these differences are essential for the overall operation of the computational artifact. Collaboration focuses our attention to the complementary piece of the puzzle—which, as Suthers aptly put it, is *the effort to maintain a joint conception.*[10] In our study, need to maintain a joint conception was evident in two forms: the pressures from sharing resources with the learner's partner and with the class, and having to decide together how to work within shared constraints. Sharing resources involved sharing construction materials (e.g., Lego bricks), and working within the constraints of limited physical space for designing the overall machine. Although these constraints did not always require learners to establish common ground, it became evident to the dyads that actions and decisions of one member could significantly impact the scope of work for the other, as well as for the overall functioning of the mathematical machine. A clearer need for collaboration—that is, coordinated, nonredundant contributions by each learner toward a shared objective—arose later in the design process, when the dyad had to design the following: (1) a ViMAP program that linked both their machines to the behavior of the virtual agent to produce a single geometric shape; and (2) a user guide that could explain to the users how to operate and coordinate the different components in order to generate their desired geometric shape.

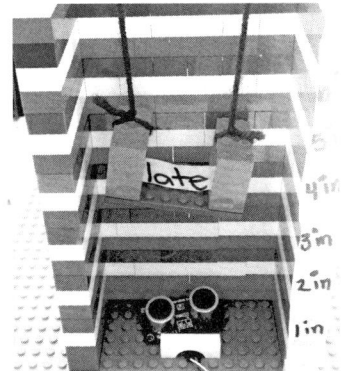

Figure 4.1
A pulley for controlling sensor readings

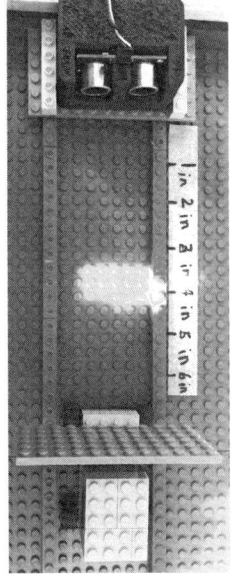

Figure 4.2
A car track for controlling sensor readings

Figures 4.1 and 4.2 show two examples of student-created machines. Figure 4.1 shows a pulley mechanism for controlling turn of the Turtle via Sensor 1. To operate this machine, a user would raise or lower the Lego plate for controlling the distance of the plate from the sensor placed at the bottom of the structure, by pulling the strings attached to the Lego plate. The labels on the right indicate vertical distance from the sensor, measured in inches. Therefore, placing the Lego plate at a distance of 6 inches would require multiplying this reading by 15 in order to rotate the Turtle by 90 degrees. This would be addressed in the ViMAP program using loops, that is, the REPEAT [number] command block. Figure 4.2 shows another example of a student-designed machine in the form of a track for racing cars. Similarly to the previous image, using this machine to move or rotate the Turtle also relies on multiplicative reasoning. The reading of the sensor is generated by moving the Lego plate (the "car") toward or away from the sensor; the label on the right indicates the distance from the sensor. The ViMAP program would need inputs from both sensors (that is, both machines) in order to generate the shape. One of these machines would control the movement of the Turtle, and the other would control the rotation of the Turtle.

4.3.2   Addressivity in Designing for Authentic Users

Understanding the needs of the users in order to design *usable* systems is the central objective of *user centered design*.[11] It emphasizes certain practices that designers must undertake in order to support understandability and usability, such as providing a good conceptual model to the user and making the controls of the system visible. That is, the system must be designed so that *during* the experience of using the system the user is able to understand how the system works, which in turn allows the user to predict the effects of their actions on the designed system, and must be able to intuitively understand how to operate and control the system.

Within the user centered design (UCD) life cycle, we focus on a subdomain of software engineering practice known as *requirements engineering*.[12] The central elements of requirements engineering are the conceptualization and development of specifications for information systems. Requirements engineering lends itself as a suitable reference discipline for this research because of the context of the pedagogical approach, which involves designing software, and its emphasis on *sensemaking*.[13] In requirements engineering, the stakeholders (that is, end users) are involved in both subjective and intersubjective interpretations about the nature and intent of the system being designed.[14] An important goal of this experience is to help the stakeholders (and designers) develop a shared understanding of the behavior of the information systems and also the underlying mechanisms and principles that govern the behavior.[15]

The focus on requirements engineering makes explicit that addressivity is also an important element of the professional practice of software engineering.

Adopting this as a pedagogical practice often frames computational design as design for instructional use and invites potential users into the design process. Previous research suggests that designing for instructional use can indeed act as a productive pedagogical approach for K–12 science and math education. For example, Brown and Campione[16] noted that 5th and 6th graders developed a deeper understanding of science concepts while they sought causal explanations to incorporate into HyperCard documents that they developed to teach their classmates. A few notable studies have also focused on children designing agent-based simulations as instructional software for mathematics[17] and instructional games in science and mathematics.[18] These studies illustrate a potential reflexivity between disciplinary practices in science and math, and addressivity in design.

However, an interesting finding across these studies is that although children find designing for use to be quite motivating, they do not consider the involvement of users as a useful component of their design process. For example, Carver and colleagues[19] found that "getting someone to try out the presentation" was regarded by middle school students as one of the least important tasks to accomplish during their design process; instead, they believed that the designers themselves could act as users during the design process. Furthermore, when children designed educational software without involving real users in their design process, they designed user interfaces that were confusing for the real users.[20]

These studies suggest that the involvement of users during the design process therefore requires explicit instructional design. In fact, once real users had tested the children's designs, Carver and colleagues[21] found that the design documents designed by the children to scaffold user interaction with their designs increased greatly in terms of making explicit the connections between the different aspects of their design and of explaining how to use the designed artifact. To this end, we show how powerful the experience of taking on the perspective of the user—especially by *being with* the user—can be, and how it can help us broaden and deepen our own understandings (as researchers and educational designers) of what it means to voice code in STEM classrooms.

4.4 The Study

4.4.1 The Setting

The study was conducted in a 4th-grade classroom of Connor Academy as part of its regular math classes. Connor Academy [22] is a predominantly (95 percent) Black public charter school in the mid-southern US. The class had 14 students (8 male and 6 female), all of whom participated in the study. For the study, we worked with the class twice a week for a total of 35 days over a period of six months (October–May). The daily lessons were co-designed and led by Ms. Lena, the classroom teacher, in collaboration with Pratim. Because this was an elementary grade classroom, Ms. Lena was responsible for teaching all curricular topics to the students. Two additional members of the research team were present during several of the classes in order to assist with data collection and to provide logistical support.

4.4.2   The Sequence of Activities

During the first eight class periods, the activities involved creating geometric shapes (e.g., squares, circles, and spirals) using ViMAP. This was motivated by Ms. Lena's intention of integrating computational modeling activities with her leading regular mathematics curriculum. This enabled her to link several curricular units in her math curriculum, such as multiplicative reasoning, geometric shapes, angles, and so on, with learning to program ViMAP. For the next 18 class periods, students worked in dyads on constructing mathematical machines and user guides.

At the very outset of the second phase, we had emphasized that "real" users would be invited to use their work. This was important because as we have previously mentioned, prior research has shown that children are generally unwilling to involve users in their design work.[23] Our goal was to position "users" as relevant others for the student designers from the beginning of their design journey. On the basis of a classroom-wide discussion at the beginning of this phase, students collectively agreed on an appropriate group of real users consisting of 4th-grade teachers in the local school district (not in their school), who would also find the machines useful in their teaching. However, due to logistical issues, we invited three graduate students in education with prior math teaching experience in elementary grades to serve as users. Some of them had served as teachers in local districts, and none of them were otherwise affiliated with the study.

The first user testing took place in mid-March (User Testing 1 or UT1), and the second user testing took place in late April (User Testing 2 or UT2). Each user testing session lasted about 3 hours. In these sessions, a user interacted with a dyad's machine for about 20–30 minutes, and gave the students written and verbal feedback. After User Testing 1, students improved their machines and user guides in order to address the issues highlighted in the feedback. User Testing 2 was also the capstone activity.

4.4.3   Data and Analysis

We collected audio and video data in the form of video and audio recordings of classroom activities and user testing sessions. We also saved copies of student-created computer programs (ViMAP programs) and the user guides, and photographs of their physical machines. In addition, we conducted and videorecorded in-depth interviews with the students throughout the year. Written field notes of the researchers were also consulted throughout the analysis.

Our analysis focuses on identifying (1) how dialogical engagements with the intended user can shape the students' experiences of computational design and can be supported through classroom instruction, and (2) how such engagements can deepen students' disciplinary learning (in this case, mathematical reasoning). This is based on our emphasis on addressivity in computational design, which both is an epistemological focus of our work and is evident in the data.

We present a qualitative analysis conducted using the constant comparative method with a focus on theoretical sampling.[24] In theoretical sampling, the search and selection of relevant data and the theoretical purpose justifying the selection are inextricably interlinked. Our theoretical focus on addressivity guided the selection of the cases for analysis, and these cases give shape to the theoretical perspectives presented in the chapter. The cases include vignettes of classroom interaction, interaction of students with the teacher and with users, and interaction between student dyads. Our selection of cases was based on identifying themes that would most vividly illustrate addressivity in computational design.

What counts as addressivity emerged both from our reading of the literature and from our interpretation of student experiences. We built on an earlier analysis[25] in which we focused exclusively on the improvements in student-generated artifacts during UT1 and UT2. With these improvements kept in mind, we reviewed videos of classroom discussions, interactions, and interviews with different groups between UT1 and UT2 and during UT2, and were able to identify several episodes that offer explanatory accounts for improvements we noticed in students' artifacts in UT2. We found that all these episodes are accounts of students making improvements in their designs through trying to make their work explicit to *others*: the classroom teacher (Ms. Lena), who would sometimes play the role of "a user," or the users themselves.

The central theme that emerged from this analysis was the shift from "playing the role of the user" to "being with the user." Comparison across the different dyads played an important role in strengthening the quality of analysis, as it helped us identify emergent subthemes that were common across the groups (e.g., improvement in user guides in UT2), as well as subthemes evident only in a few groups (e.g., the reframing of Ms. Lena as a "user" for Jacinda and Tom). The vignettes we have selected represent both these forms of subthemes and also illustrate how this shift shaped learners' experiences of computational design and deepened their STEM disciplinary learning (e.g., mathematical reasoning).

4.5   Finding: The Shift from Being the User to Being "with" the User

Design is a reflective conversation with the situation, which also involves taking the perspective of the user. One could argue that each student acted as the user every time they tested their machines and their code. This is akin to the students projecting themselves onto the user, and it was really evident during their interactions with the users in UT1. During UT1, we noticed that the student-designers took over control of the machines from the users by physically removing users' hands from the machines and the computer and proceeded to perform the activity on behalf of the users. We found this to be the case especially when the users would ask them to explain how the program was linked to the mechanical operations or how their actions generated the shape. In contrast, we found that during UT2, in all groups, there was an overall ceding of control of the machine and the code to the user, and the student-designers would intervene only upon request of the user.

Let us now look at an illustrative case of Daniel and Jason, which will make explicit the shift between the two experiences. This case serves as a powerful illustration of the shift in students' designs, which was also borne in their designed artifacts. To this end, we also present comparisons between their UT1 and UT2 designs. In addition, we present vignettes of classroom interaction that occurred between UT1 and UT2, which illustrate how the teacher supported this shift.

4.5.1   Being the User (UT1)

During the first round of user testing, we noticed a pattern that was common across all groups: students took over the control of the machine and the computer on several occasions. This would happen quite often as the users were unable to follow the instructions in their user guides (a common occurrence), as well as when the users would struggle with accomplishing a particular objective. Both these forms of trouble were evident in Daniel and Jason's group.

Their user, Faith, was a mathematics teacher who was also enrolled in a master's program in education at the local university. Faith wanted to create her own program in ViMAP from scratch, and soon realized that she did not find the relevant instructions in the user guide. Jason literally took over the control of the laptop from her and dragged a couple of blocks from the library of commands onto the programming palette. Faith commented that it would be beneficial if there were an explanation of how to create programs in ViMAP by dragging and arranging blocks. He initially dragged the *pen down* block, and then dragged the following block into the palette: *set ((STEP-SIZE)) ((SENSOR 1)) times ((10))*. Upon sensing that Jason was proceeding to create the entire program, Faith requested him to let her try to do so first, and then Daniel and Jason could advise her on every step.

Faith asks Jason to explain why she would have to use the block:

set ((STEP-SIZE)) ((SENSOR 1)) times ((10))

and then tries to move the block out of the programming palette. Jason replies:

Jason: You can't . . . you gotta use that block . . .

Upon being prompted, Jason offers no further explanations. Faith then asks him to point to Sensor 1. Jason replies:

Jason: Everytime we do it we get confused—so we just test it.

Jason then starts moving the cars to explain which sensor controls the turning and which one controls the step size.

This episode illustrates that Jason and his partner were more comfortable with demonstrating their machine than letting the user try it out. Because they had tested their machine and their code themselves while playing the role of the user, many of the questions that arose during Faith's interactions with the machine seemed familiar to them, but it was clear that they had not realized how to resolve these issues for others. Guessing what a user (unknown to the designer) may find relevant or important to know or notice is a nontrivial issue, and it is specifically for this reason that scholarship on requirements engineering has identified involvement of the user during the design process.[26] On the other hand, taking over control of the machine from the user put themselves in charge, an already familiar situation for them.

Therefore, this simple episode, representative of all the groups in the class during UT1, is a powerful illustration of the student-designers being limited to testing their own designs as users. The explanations that Faith asked for were not merely informational, but involved significant mechanistic reasoning. During the debriefing after the UT1, the teacher noted to Pratim that she realized the learning opportunities inherent in creating such explanations. In the next section, we see how she acknowledged these difficulties as relevant to learning by leading a productive class discussion, and also made explicit how the different forms of trouble experienced by the users could be helpful for refining students' work.

4.5.2   The Heterogeneity of Trouble

It is now late February, and the 4th-grade students at Connor Academy (CA) are asking us about the pizza party that we have promised them in celebration of their hard work on designing the machines after the first session of user testing. Right before the class, students are trickling into the classroom in groups of two or three from recess, and we also hear them talk about their new schools next year (students will disperse to middle schools for their 5th-grade year). But today, they are particularly vocal about what happened two days earlier. "Users" from outside their class—teachers and mathematics education graduate students from a local university—visited their classroom and tested their machines.

However, as we begin talking with the students, we realize that the students are not happy with their experience of how the users interacted with their machines, or with the questions that the users had asked them. As the teacher, Ms. Lena, walks in, she also tells Pratim that she needs to lead a class discussion so that the students can discuss *their* experience of user testing. "There was a mutiny here yesterday," Ms. Lena tells Pratim, "Students do not want to work on the designs anymore; they think the users did not understand much."

The frustration is palpable. Ezekiel, who had been confident about his design up until now, is sitting down with his head on his desk. He looks up as Pratim walks by and tells Pratim, "Users are trouble. *Reaaal* trouble." The students sitting around him nod in agreement, and several students join Ezekiel in voicing their disapproval of the users.

Epistemologically, this is a significant moment, as it can help us understand the difference between *imagining* the user and *seeing* the user, especially from the perspective of the student-designers. The notions of "use" and the "user" are inherent in our understanding of design. While these notions are often implicit in our experience as designers, the *actual* experience of interacting with users bears significant differences. This was also true for the students, who by now had been working on their machines and ViMAP programs for three months, twice a week. Sustaining a long group project that spans several months (as opposed to one to two class periods) was challenging for most groups. Five out of the eight groups reported problems pertaining to team dynamics within their groups, which included issues such as sharing resources and misalignment of design objectives. However, in each case, the teacher would remind the group that they were responsible for designing a "product" that will be used by "real users," which in turn served as the motivation to resolve these differences.

In contrast, "seeing the users," which involved meeting them, being with them, interacting with them, and listening to them, was a remarkably different experience. What the users actually did stood in contrast with what the students had imagined they would do. The class discussion that Ms. Lena led revealed that all the students, in their roles as designers, felt challenged by the users' interactions with the machines and the code, as well as their questions. In particular, students noted that they found the users' questions about how each mechanical action on the machine was related to mathematical ideas to be quite difficult. They also noted that even when they thought that they had explained the mathematical relationships between the ViMAP code, the geometrical shapes that were being "drawn" by the ViMAP Turtle on-screen, and the relevant components and actions of the physical machines, the users did not seem to understand their user guides, and were confused about how to use the machines.

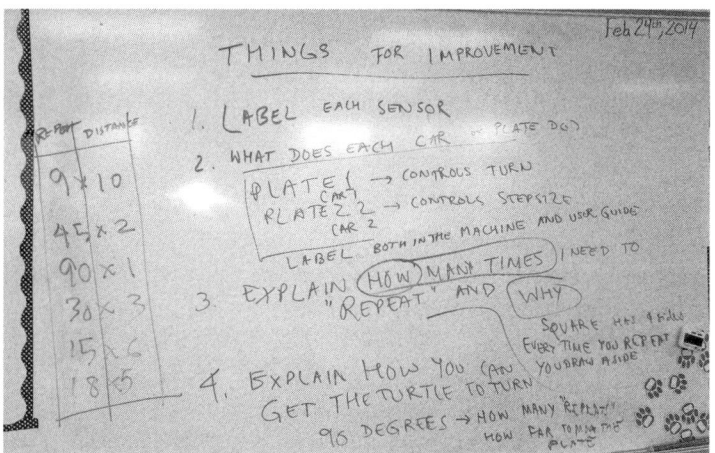

Figure 4.3
Students and teacher identify the types of improvement after first user test, and Pratim records key ideas on whiteboard

Each of these kinds of trouble was noted by the teacher as important to work on. Ms. Lena asked Pratim to help her with the class discussion by being the scribe on the classroom whiteboard. She reminded the class that the users would be invited back again in a few weeks once the class had an opportunity to revise their machines and the user guides on the basis of the users' feedback. As students began to share the issues they had noticed during user testing, Ms.

Lena asked them to think about what to change in their designs, including their machines, code, and user guides, in order to address these issues.

Figure 4.3 shows the themes that emerged from the resulting class discussion. It captures eloquently the heterogeneity in terms of the different elements of the designed artifact that needed further refinement, as well as the intertwined nature of these elements. Paying attention to these heterogeneous facets, in turn, implied disciplinary heterogeneity: the explanation of their ViMAP code was inextricably tied to the underlying mathematical reasoning, and the explanation of how each mechanical component of the mathematical machine operates was also tied to the explanation of the ViMAP commands and the emergent mathematical shape. For example, multiplicative reasoning is evident in the table on the left that lists different pairs of numerical values, the product of which is 90. This value represents the angle between adjacent sides of a square; the left column represents the number of loops in the ViMAP code (labeled "Repeat"), and the right column represents the distance between the relevant sensor and the Lego plate (labeled "Distance"). The need to create communicative designs is evident in each of the four points in Figure 4.3: Labeling each sensor to make the user aware of the different parts of the machine ("Label each sensor"); explaining explicitly how the user needs to operate the machines ("What does each car or plate do?"); making explicit the underlying multiplicative rules needed for operating the machine ("Explain how many times I need to repeat and why"); and explaining how to get the ViMAP Turtle to "turn by 90 degrees."

4.5.3   Being *with* the User (UT2)

During UT2, we saw a different image of Jason and Daniel. At the beginning of this interaction, Faith asks them to explain how she can draw a square. Daniel and Jason proceed to explain how each car affects the movement of the Turtle (Daniel) and how the markings on the track correspond to the different elements of the program (loops, turn angles, and number of sides). Faith then decides to take this a step further, and this is where the excerpt reported here begins. She ask them if she can "have fun" and "mess around" with the machine. In the transcript below, simultaneous speech is written in parallel columns, and (parentheses) indicate pointing or signifying gestures.

> **Turn 1** Faith: Can we now do something fun?
> **Turn 2** Daniel: Uh
> **Turn 3** Faith: Can we mess around with this *(points to the left and right several times, looking at the machine)*?
> **Turn 4** Jason: Yes! *[at the same time]* Daniel: *(nods)*
> **Turn 5** Faith: What can I—what can I mess around with?

> **Turn 6** Daniel: Uh, everything *[points at machine; Jason nods in agreement]*.
> **Turn 7** Jason: You can start with this one. *[Jason leans over and rolls one car back.]*

Jason and Daniel were not prepared for this question, but are eager to participate. This is perhaps most strongly evident when Daniel tells Faith that she can mess with "everything" (Turn 6: see excerpt above), and Jason directs Faith to one of the cars as a place to start "messing" with. Faith is about to move the cars, but Ms. Lena signals that the class is running out of time. Faith is still eager to know what will happen if the cars are moved, and sensing this, Pratim, who was recording the interaction, asks Jason to explain the effect his change of moving the car will have on the drawn shape. Both students answer, as the following excerpt shows:

> **Turn 8** Pratim: What will happen, do you think?
> **Turn 9** Faith: Wait wait wait . . .
> **Turn 10** Pratim: What, Jason, what will happen?
> **Turn 11** Daniel: It will get small—it will mess up the, uh, square.
> **Turn 12** Pratim: It will mess up the square. How will it mess up the square?
> **Turn 13** Daniel: It will probably make the uh, it probably will widen out.
> **Turn 14** Jason: I—I—I'll move mine back so the right angles—the square go out which is—when it closes in, it closes.
> **Turn 15** Pratim: Alright.

When students explored outside the boundaries of their previously prepared materials, the gesture helped to represent their new ideas when other parts of their representation became inadequate or irrelevant to the task. Daniel and Jason are now *voicing* code using *representational gestures*.[27] Daniel enacted the gesture of the shape "closing in" (Figure 4.4), and Jason enacted the gesture of the shape "widening out" (Figure 4.5). After Faith left the group, Pratim asked Daniel and Jason why they chose not to use the machine to "demonstrate" their answer. Jason explained that doing so would be time-consuming, as well as require them to take control of the machine away from the user. Taking control away from the user was something that was discouraged as part of the preparation for UT2. Jason further explained that *"when we showed (Jason gestures to refer to his use of gestures during his interaction with Faith) how it worked, she got it."*

The gestural explanation served the purpose of *simulating* the relationship between movements of parts of their machine and how the shape would be affected by it. Although the user could not yet see the change in the shape, and due to time constraints neither the students nor the user were able to physically create the new scenario, the gesture filled this visual and experiential gap. The students used their bodies to represent the mathematical relationship between sensors and code, thus making explicit a mechanism hidden from the user's plain sight. In addition, similar to Becvar and colleagues,[28] gestures also served the purpose of a seamless and efficient means of communication by adhering to one of the key requirements of the task (not taking control away from the user), while representing and communicating a complex conceptual relationship in a collective setting.

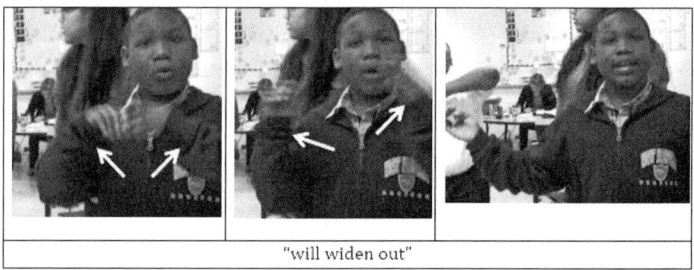

Figure 4.4
Daniel's gesture for "widening out" in Turn 13

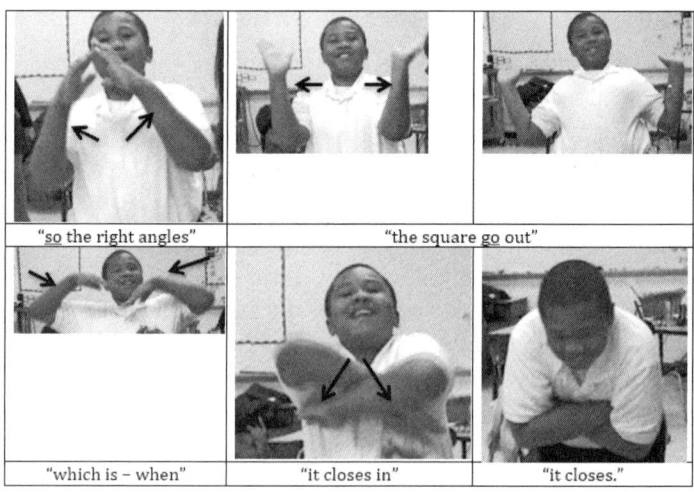

Figure 4.5
Jason's gesture for "closing in" in Turn 13

4.5.4   User Guides as Computational Utterances

A comparison between students' user guides in UT1 and UT2 revealed that in User Testing 2, the mechanical components of their machines and their user guides were designed in order to make explicit to the user the relationships between their (the users') interactions with the machine and the computational output of the ViMAP program. We found that all the groups accomplished this by creating discrete mathematical explanations corresponding to different sets of user actions. For example, as shown in Figure 4.7 and as we explain later in this section, the explicit linking of discrete mathematical explanations to user actions (e.g., users placing their hands at a certain distance from the sensors) was illustrative of the shift from a monologic stance in UT1 to a dialogical stance in UT2. In UT1, students typically listed actions with minimal explanations and they also lacked specificity in terms of what exactly users needed to do. In contrast, user guides in UT2 could be seen as computational utterances, because they were addressed to users and designed through experiences of being with the users.

One role that this form of addressivity played was deepening the nature of mechanistic explanations, as evident in students' user guides and the structural improvements in the machines in UT2 over those in UT1. For the readers' convenience, here we present a brief recount of a more detailed analysis of mechanistic reasoning as evident in students' artifacts, which we have published elsewhere.[29] Mechanistic reasoning involves more than simply mentioning that X causes Y to happen; it requires that students come to identify how X brings about Y.[30] Comparison of the UT1 and UT2 user guides revealed a general trend across all groups: the earlier version of the user guides specified user actions, but did not specify how the actions were related to the desired computational output. In contrast, the later iteration of the user guides explained relationships between user actions, elements of the ViMAP program, and the component Turtle behaviors that resulted in the desired shape.

An illustrative example is the case of Jacinda and Tom. Jacinda and Tom designed a two-car setup (Figure 4.8), consisting of two separate, manually operated, wheeled cars. Each car had a flat surface (a wall made of Lego bricks) attached to its front, representing the palm of a hand. Pushing each car toward or away from the sensor generated a reading of the ultrasonic sensor, based on the recorded distance of the Lego wall from the sensor. The distance of one of the cars from the sensor controlled the speed of the ViMAP Turtle, and the distance of the other car from the second sensor controlled the rotation of the Turtle.

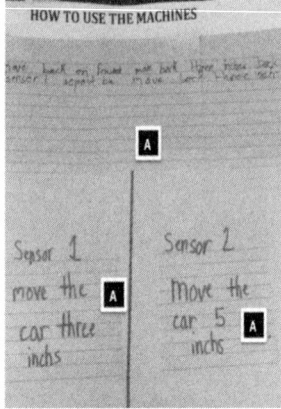

Figure 4.6
Jacinda and Tom's user guide in User Testing 1

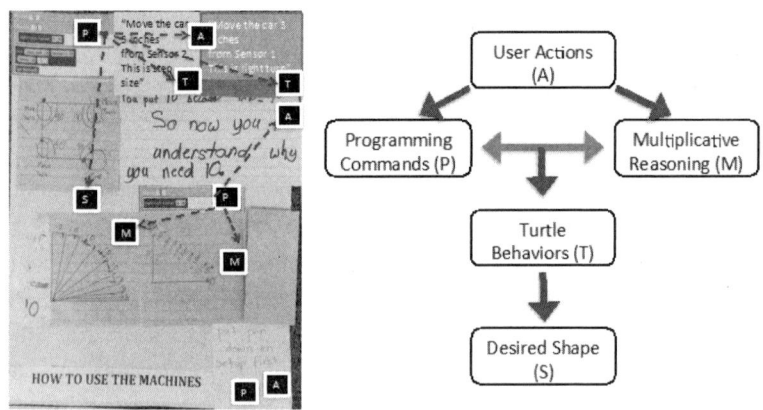

Figure 4.7
Left, Jacinda and Tom's user guide in User Testing 2 and, *right,* schematic used for analysis of mechanistic reasoning

Figure 4.8
Jacinda and Tom's machine in User Testing 2

In their user guide for UT1 (see Figure 4.6), Jacinda and Tom adopted a monologic stance. They simply wrote down instructions that specified the distance from the sensors at which the user would have to position the two cars in order to generate specific readings. In contrast, in UT2 their user guide contained diagrams and written explanations intended to explain the multiplicative reasoning, which in turn explained how actions of the user were related to specific commands and parameters used in their ViMAP program (Figure 4.7). For example, in one of the drawings in their user guide, Jacinda and Tom explained that turning 90 degrees involved repeating a turn of 10 degrees (generated by the reading on Sensor 1) nine times. Adjacent to the drawing was a cutout of a color printout of the corresponding ViMAP program, in which the turn angle was repeated nine times using the "Repeat [number]" command. In their verbal explanations during UT2, both the students were able to point to the relevant parts of their user guide and provide similar mechanistic explanations. That is, we found that their user guide and verbal explanations during UT2 used multiplicative reasoning to explain the following: (1) how user actions generated Turtle behaviors (such as movement and turning), and (2) how the overall shape was generated by repeating these actions computationally using programming loops.

On the basis of the feedback in UT1, Jacinda and Tom also made improvements to their machine that demonstrated a dialogical stance. They implemented structural improvements so that users could operate their machine more easily. In UT1, users faced difficulties in moving the cars due to an uneven track; the cars lacked stability and would topple while moving. In addition, users experienced difficulties in translating the "actions" specified in the user guide to interactions with the machine. To address these issues, Jacinda and Tom flattened the surface of the track on which users would move the cars, and strengthened the base of the cars to prevent them from toppling. They also labeled the sensors and different parts of the track in ways that complemented and supported the mathematical explanations in their user guide. That is, their labels placed on specific parts of their machine were designed to inform the user about what to do with that part and also about the mathematical justification for it (Figure 4.8). The labels specified the number of programming loops that the user would need to put in their ViMAP program to generate a 90 degree turn, corresponding to three different positions of the car that controlled the turn angle of the Turtle. Five out of eight groups designed such explanatory labels on their machines in UT2, whereas none of the groups designed such labels in UT1.

In reviewing the video recordings of the interactions that Ms. Lena had with individual groups between UT1 and UT2, we found a short but revealing episode. This excerpt represents a brief conversation that Pratim had with Jacinda and Tom that took place immediately after they had shown their user guide to Ms. Lena. This was several days past UT1, and Jacinda and Tom had already made several changes to their user guide on the basis of the class discussion reported in section 4.2 as well as their own experiences of UT1. Their user guide at this point had the explanatory diagram shown on the bottom right of Figure 4.7, but not the one to the left of it. It was only after this interaction that Jacinda and Tom created the representation on the bottom left in order to make even more explicit that nine rotations of 10 degrees would result in a total rotation of 90 degrees. This excerpt is shown below.

> **Turn 1** Pratim: What happened?
> **Turn 2** Jacinda: What happened is that I got a user and she did not understand what this is. . .*[points to the diagrammatic* xy *representation of a right angle along with nine dots spaced diagonally about 10 degrees apart]* so I am gonna change it by changing it right there and making an arrow right there. *[Jacinda then points to the empty space to the left of the diagram]*
> **Turn 3** Pratim: . . . And who was your user?
> **Turn 4** Jacinda: Ms. Lena *(the teacher).*

The most notable thing here, however, is that Jacinda refers to Ms. Lena as a user. This is a significant shift away from the preparation for UT1, in which each group was encouraged by the teacher to "think" and "act" like the user. We also heard this from two other groups, who started referring to Ms. Lena as the "user." In absence of the user, Ms. Lena decided to embody the user, as she noted to Pratim after this conversation, because in letting her interpret their user guides the students would be able to engage in a conversation about what works and what does not. This is how the dialogical frame was kept alive between UT1 and UT2 by the teacher.

4.6 Epilogue: Publicness and the Computational Utterance

We began this chapter with a critique of the individualistic focus in computing education. We argued that a heteroglossic framing can orient our attention to an essential publicness in the experience of computational design. We positioned Bakhtin's notion of addressivity—the constant state of being addressed and being in the process of answering—as central to the dialogical imagination of computational design. In enlivening addressivity, we focused our attention on making the addressee (the user) visible in the students' experiences of computational design.

Foregrounding addressivity positions computational media design as a search for otherness. Making the others explicit led to representational complementarity and heterogeneity, which then necessitated students' deepening their mathematical explanations. These explanations in turn served as transformations and translations between representational genres and systems. This is how dialogicality shaped students' disciplinary learning as well as their design experience. Their explanations made explicit the mathematical relationships between algorithmic elements (e.g., number of loops in their ViMAP program) and the actions of the Turtle in every step (e.g., right turn), which in turn was directly affected by the users' actions (e.g., sensor reading generated by the user). We consider this as evidence of the reflexivity between user-centered engineering design and mathematical learning.

The publicness of the experiences reported in this chapter lies in the worldliness of code and the computational utterances. Enlivening alterity and addressivity played a key role and helped the student-designers *see* their work as belonging both to themselves and to the world (the users). The continued (iterative) involvement of the user and the teacher's role-playing as the user are examples of such images of alterity and addressivity *in action*. This is also resonant with studies of professional practice of software design, which have shown the involvement of the user to be important for articulating design requirements and shaping the discourse around design.[31]

Through emphasizing addressivity, we also recast individualistic and device-centered notions of ownership and mastery in a fundamentally dialogical frame. This is a heteroglossic imagination in which agency in computational design is repositioned as *being present with users*, not merely predicting the actions of hypothetical users. Our work shows that such an approach can untether the student-designers' experiences of computational design from computational artifacts in a productive manner and that their artifacts become more vividly addressed to the users. It is through such experiences of voicing code *for and with others* that a functional computational artifact becomes a computational utterance in the context of computational design.

Notes

1. S. Papert (1980). *Mindstorms*. Basic Books.
2. A. A. diSessa (2001). *Changing minds: Computers, learning, and literacy*. Cambridge: MIT Press, 83.
3. M. G. Ames (2019). *The charisma machine: The life, death, and legacy of One Laptop per Child*, 46. Infrastructures. MIT Press.
4. See section 1.2 for a detailed discussion.
5. M. Heidegger (1962). *Being and time*, 117–125. Harper & Row.
6. Heidegger, *Being and time*, 1962, p. 122.
7. L. Steinby & T. Klapuri, eds. (2014). *Bakhtin and his others: (Inter)subjectivity, chronotope, dialogism*. Anthem Press.
8. D. D. Suthers (2006). Technology affordances for intersubjective meaning making: A research agenda for CSCL. *International Journal of Computer-Supported Collaborative Learning*, 1(3), 315–337.
9. S. Papert (1980). *Mindstorms,* 187–194. Basic Books.
10. Suthers, Technology affordances, 2006, 318. See also: J. Roschelle & S. D. Teasley (1995). The construction of shared knowledge in collaborative problem solving. In *Computer supported collaborative learning*, 69–97. Springer.
11. D. A. Norman & S. W. Draper (1986). User centered design. *New Perspectives on Human-Computer Interaction*, 3161. L. Erlbaum Associates.
12. S. Chakraborty, S. Sarker, & S. Sarker (2010). An exploration into the process of requirements elicitation: A grounded approach. *Journal of the Association for Information Systems*, 11(4), 212–249.
13. P. Ovaska & L. Stapleton (2009). Requirements engineering during complex ISD: A sense-making approach. In *Information systems development*, 195–209. Springer.
14. S. Chakraborty, C. Rosenkranz, & J. Dehlinger (2015). Getting to the shalls: Facilitating sensemaking in requirements engineering. *ACM Transactions on Management Information Systems (TMIS)*, 5(3), 14. See also: C. Rolland (1994). Modelling the requirements engineering process. In *Proceedings of the 3rd European-Japanese Seminar on Information Modelling and Knowledge Bases*, IOS Press, 6.
15. Chakraborty et al., Getting to the shalls, 2010.
16. A. L. Brown & J. C. Campione (1990). Communities of learning and thinking, or a context by any other name. *Contributions to Human Development,* 21, 108–126.
17. I. Harel (1990). Children as software designers: A constructionist approach for learning mathematics. *Journal of Mathematical Behavior* 9(1), 4.
18. Y. B. Kafai, M. Franke, C. C. Ching, & J. Shih (1998). Game design as an interactive learning environment fostering students' and teachers' mathematical inquiry. *International Journal of Computers for Mathematical Learning,* 3(2), 149–184.
19. S.M. Carver, R. Lehrer, T. Connell, & J. Erickson (1992). Learning by hypermedia design: Issues of assessment and implementation. *Educational Psychologist* 27(3), 385–404.
20. Kafai et al., Game design, 1998.
21. Carver et al., Learning by hypermedia design, 1992.
22. Pseudonym.
23. Kafai et al., Game design, 1998; and Carver et al., Learning by hypermedia design, 1992.

24. B. G. Glaser & A. L. Strauss (2017). Theoretical sampling. In *Sociological methods*, edited by N. Denzin, 105–114. Routledge.

25. P. Sengupta, G. Krishnan, M. Wright, & C. Ghassoul (2014, April). Mathematical machines and integrated stem: An intersubjective constructionist approach. In *International Conference on Computer Supported Education*, 272–288. Springer.

26. S. Chakraborty, S. Sarker, & S. Sarker (2010). An exploration into the process of requirements elicitation: A grounded approach. *Journal of the Association for Information Systems*, 11(4), 1.

27. M. W. Alibali & M. J. Nathan (2012). Embodiment in mathematics teaching and learning: Evidence from learners' and teachers' gestures. *Journal of the Learning Sciences*, 21(2), 247–286.

28. A. Becvar, J. Hollan, & E. Hutchins (2008). Representational gestures as cognitive artifacts for developing theories in a scientific laboratory. In *Resources, co-evolution and artifacts*, 117–143. Springer.

29. Sengupta et al., Mathematical machines, 2014.

30. R. Russ, R. Scherr, D. Hammer & J. Mikeska (2008). Recognizing mechanistic reasoning in student scientific inquiry: A framework for discourse analysis developed from philosophy of science. *Science Education*, 93, 875–891.

31. Chakraborty et al., Getting to the shalls, 2010.

5 Recontextualization and Transitional Othering

5.1 Sociopolitical Emergence and the Need for a Critical Computing Education

There is now a growing acknowledgment that direct engagement with social justice must become a part of technoscientific literacies.[1] Although it may be tempting to see such efforts as merely new contexts for embedding technoscientific work, taking into account issues of power, access, and inequality can actually *deepen* disciplinary inquiry.[2] In the context of supporting computational and data literacies, Philip and colleagues argue that shielding STEM classrooms from "racially charged examples" is counterproductive, given the expectation that students will eventually engage in STEM-based reasoning as active citizens.[3] The synthesis of STEM disciplinary knowledge and practices on one hand, and civic, socioscientific, and sociopolitical issues on the other, is challenging and does not happen automatically.[4] Therefore, engagement with the social and historical complexities that are deeply interwoven in socioscientific data and data visualizations must be deliberate, rather than accidental. Using a neighborhood-wise heatmap of movie choices in the city of Los Angeles, Philip and colleagues showed that conversations about race, power, class, and inequality can deepen students' interpretations and understandings of the data.[5]

Computational models, however, go beyond data visualizations of race. This is an important recognition especially in light of recent arguments that have posed a fundamental challenge to the notions of technological neutrality.[6] The growing recognition that racism and racial bias may also be deeply entrenched within the built infrastructure of computing—programming languages,[7] hardware,[8] algorithms,[9] and data[10]—makes it imminently important for educators and students to engage directly with inner infrastructure of computing. Furthermore, modeling race computationally is a far from settled issue, because the universalist nature of computational abstractions may not be powerful enough to represent the local, embodied, and affective nature of experiences of racialization, and more generally, intersectional forms of oppression and marginalization.[11]

Therefore, the question of what computational representations of race and racialization should highlight—and, to what end (that is, in terms of desired learning experiences)—are nontrivial ones, but questions that we need to ask urgently, both in contexts of computing in practice and society, and K–12 education. These questions go to the heart of what counts as (and should count as) *authenticity* in STEM and computing education. This is particularly important because as Philip and Sengupta recently noted,[12] notions of authenticity in computing education has largely ignored systemic oppression of people of color. Our chapter is an invitation to the field—particularly computing and STEM education scholars focused on K–12 contexts—to engage in this conversation. Furthermore, it has also been argued that engaging with issues such as race and power in the classroom is important for K–12 students and teachers, particularly in the US, because the gap between the cultural backgrounds of teachers (who are predominantly White) and students of color is increasingly getting wider.[13]

In this chapter, we present an account of how computational abstractions can be *voiced* in the form of discourse about sociopolitical and historical inequalities by preservice teachers. We explore how a multiagent simulation of ethnocentrism, despite its inevitable shortcomings, can be used with preservice teachers to support critical conversations about racial segregation. Our work is premised on designing pedagogies around a particular form of computing (multiagent-based computing) that has been shown to lend itself well to modeling complex, emergent systems in ways that are also approachable and inviting for students[14] and preservice teachers[15] with no previous programming experience. In emergent systems, collective or system-level patterns *emerge* from interactions between many individual actors. Using multiagent-based computational models, students and preservice teachers can create and modify individual-level rules that govern the behavior of these individual actors (computational agents), and can interpret and understand how even counter-intuitive collective-level patterns can emerge from interactions based on these simple rules. We illustrate how multiagent-based computational representations can be leveraged to create an inviting environment for students and teachers to engage in difficult and complex conversations about urban segregation and inequality.

This chapter extends and deepens our previously published work,[16] in which we showed that the use of computational agents as conversational scaffolds supported teacher candidates' engagement in these conversations, although initially they were unwilling to engage in such conversations without the scaffolds. Here we dive deeper into a more fundamental form of epistemological analysis that illustrates how computational abstractions and representations become sociopolitically meaningful through progressive *recontextualization*. To accomplish this, we present two forms of analysis: (1) an analysis of published research in computational social sciences on multiagent simulations of ethnocentrism, and (2) an analysis of how preservice teachers begin to engage in critical conversations about race and inequality as they begin to recontextualize the multiagent model using complementary representations, through a form of experience that we term *transitional othering*.

5.2 What Does Ethnocentrism Model?

It is important to note that we do not refer to ethnocentrism as a social phenomenon, but as a computational model that is widely used in the social sciences and in computational science. However, as we mentioned in chapter 2, a model only amplifies certain aspects of the phenomenon it represents, while hiding and even omitting others. We will go deeper into the computational representations in section 5.3. In this section, we explore some of the amplifications and omissions of ethnocentrism as a representation of race in society, rather than positioning the use of simulations of ethnocentrism as the most appropriate way to engage in critical conversations about racial segregation.

5.2.1 Amplifications and Omissions

Ethnocentrism was coined as a construct by Western sociologists in order to study in-group formation and behaviors.[17] Simply put, the term indicates in-group favoritism at the expense of other groups. Theories and models of ethnocentrism have been used to study and explain a wide range of phenomena such as intergroup bias,[18] consumer behavior,[19] racial dynamics,[20] and others.

Computational approaches using multiagent simulations explain ethnocentrism in terms of evolution of strategies, and collective-level patterns of ethnocentric behaviors as emergent phenomena.[21] The agents in these simulations can either cooperate with, and/or not cooperate with, in-group or out-group agents. This generates four possible strategies: (a) a selfish or "egoist" strategy not cooperating with anyone, (b) a "cosmopolitan" strategy of cooperation with out-group, but not in-group, agents, (c) an "ethnocentric" strategy of cooperation within one's own group but not with agents from different groups, and (d) a "humanitarian" strategy of cooperation with everyone.[22] A striking contribution of this work is demonstrating how the dominance of ethnocentric behaviors at the collective-level can *emerge* through local interactions between individual agents, even when there is no initial systemic bias toward ethnocentrism.

It would be tempting to assume that we are positioning ethnocentrism as a model that can account for systemic racism. That would be incorrect, because systemic racism goes far beyond the emergence of "macro-behaviors" such as collective-level segregation from "micro-motives,"[23] and the phenomenology of race and racialization is far more complex than universalist explanations of in-group favoritism that underlie ethnocentric models. Our experiences of race and racialization encompass embodied, systemic, interactional, historical, and affective realms. For example, scholars have shown that we experience racialized emotions in our everyday interactions with others,[24] through institutional and governmental policies and macropolitics,[25] as well as intergenerational narratives.[26] In addition, the deeply troubling history of racial oppression of colonization and slavery are also not considered or included in models and theories of ethnocentrism,[27] and neither are structural and institutional factors such as explicitly racist "red-lining" policies adopted by banks that have directly contributed to residential segregation.[28]

Why then, despite the omissions highlighted above, do we still see value in engaging with simulations of ethnocentrism? Perhaps most importantly, if we want to rectify these omissions and work toward making computing and computing education anti-racist, we must engage directly with the inner sanctum of computing—code, algorithms, data representations, and models—rather than only making "mile high" critiques about technology without direct engagement with technology. Furthermore, given the popularity and influence of computational models of ethnocentrism in social science, we expect them to be adopted by education researchers seeking to integrate computing with socioscientific issues. But beyond the discourse of popularity, there are deeper, epistemological issues that motivated our engagement, which we explain in the following section.

5.2.2   Learning with Flawed Models

In developing any computational model—however contextualized it may eventually need to be—the use of symbolic generalizations is unavoidable. From a phenomenological perspective, one could argue that it is this unavoidable element of the experience of computing that underlies Wing's emphasis of "layers of abstractions" as central to computational thinking.[29] In building a computational model, a modeler may be directly using a more contextualized layer of domain-specific programming and modeling commands, which in turn are built upon another layer of more generalized programming commands and data types. The connections between these layers both limit and amplify the representational and epistemic possibilities and are further complicated through the interpretive experience of the modeler. The possibilities and flaws of any computational model are tied to the connections between these layers and how these connections represent the phenomena being modeled. From a phenomenological perspective, we believe that the experience of *recontextualization* plays a fundamental role in forming these connections, as well as interpreting how they represent the phenomena being modeled.

This chapter orients our attention to the importance of *recontextualization* of universalist abstractions as *essential* for modeling sociopolitical issues. Our work builds on recent research in computer science which suggests that generalized computational abstractions (e.g., decision support algorithms), without appropriate interpretive and semantic support, often end up worsening systemic injustice against marginalized people.[30] The core idea here is that computational models are always going to be flawed, especially in contexts that involve sociopolitical issues, but can be used productively through appropriate forms of recontextualization. This involves engaging in progressively richer forms of contextualization of computational representations as well as valuing discourse and interpretive experience. We will see an illustrative example of the significance of recontextualization in the professional work of computational scientists in section 5.4.1.

From a pedagogical perspective, education researchers have also identified the importance of contextualization of computing within STEM disciplines, noting that simply introducing a general-purpose programming language in a science classroom may make teaching and learning both science and computing more challenging.[31] This important insight has led to the design of recontextualized computational tools for K–12 STEM classrooms. For example, while NetLogo is a more generalized programming language, ViMAP, which uses NetLogo as its underlying simulation engine, was designed as a more contextualized programming language specifically for K–12 STEM education.[32] Our work here, then, also stems from the concern that recontextualization of computational abstractions is equally important for computationally representing sociopolitical issues in K–12 settings.

To this end, at the most general level, in this chapter, *we position the experience of computational abstractions, fundamentally, as recontextualization.* We show that it is only through recontextualization that the computational rules and the simulation of ethnocentrism become progressively meaningful. This is essential for addressing the unavoidable interpretive gaps in using symbolic generalizations (e.g., computational rules) as well as generalized conceptual representations (e.g., ethnocentrism) for representing our personal experiences of race and racialization, as well as the complex dynamics inherent in systemic injustice. This approach positions *seeing* and *voicing* computational abstractions, such as computational models of ethnocentrism, in relationship with our lived experiences and relevant contextualized and historical data at the heart of challenging technocentrism. In the absence of such supports, we may fall prey to the charisma of reductionist, computational models of race and begin to see the world through reductionist lenses.

Second, our analysis establishes the notion of *transitional othering* as a powerful form of pedagogical experience. We explain this notion in a more contextualized form in section 5.4.2. We believe that this is a powerful form of pedagogical experience, through which learners can both project their lived experiences of race and inequality on computational actors and, at the same time, keep these projections at a safe distance that enables them to engage in such critical conversations. This is particularly important given our overarching framing of computing itself as a heterogeneous language (see chapter 2), in which transparency plays an important role. Transitional othering is a particular phenomenological reframing of transparency that is inherent in computational languages that can be helpful as we engage in making computing education more critical and anti-racist. It offers us an account in which the layers of computational abstractions—the simulation and its underlying rules—become recontextualized through the use of complementary and contextualized representations. Through this experience, learners can begin to "see" themselves in the computational representations, but in a way that allows them enough space to discuss complex and critical issues such as race and inequality, without feeling personally implicated in ways that might hinder such conversations.

5.3   Deeper into Modeling Ethnocentrism

5.3.1   Abstractions in Coding Ethnocentrism

How can a computer interpret ethnocentrism? This is the question that we will explore in this section. Abstractions, in this section, means ways of representing ideas so that computers can understand them. We focus on an implementation of Axelrod and Hammond's model[33] in the NetLogo programming language and modeling platform, originally created by the eminent scholar of complexity, epistemology, and computer science, Uri Wilensky, who is also the creator of NetLogo. Figure 5.1 shows the NetLogo code that specifies the *key* behaviors of each individual agent.

In this model, the individuals belong to a few different groups: egoists, altruists, ethnocentrists, and cosmopolitans. Each individual (agent) has three variables: (1) color, (2) cooperativeness with agents of the same color, and (3) cooperativeness with agents of different colors. An ethnocentric agent cooperates only with agents of the same color; a cosmopolitan agent cooperates only with different agents of different colors; an altruist or a humanitarian agent cooperates with all agents; and an egoist or a selfish agent cooperates with no one.

There are several forms of computational and mathematical abstractions used in the code shown in Figure 5.1. At the very outset, we must understand that each NetLogo agent is essentially a Logo Turtle. Each agent, in turn, is an anthropomorphic representation of a person who ascribes to a particular form of strategy for interactions with others. Spatially, the *neighborhood* of each agent is defined in the form of a Von Neumann neighborhood, a mathematical formalism commonly used in cellular automation (CA) models in computer science and in statistical physics. A Von Neumann neighborhood is a two-dimensional square lattice composed of a central cell and its four adjacent cells, as shown in Figure 5.2. This means that each agent in the simulation responds only to the agents in the cells adjacent to it. This also means that changing the nature of this neighborhood would also alter the nature of interactions of each agent and, potentially, the emergent outcomes of the simulation.

Computationally, the behavior of each agent is defined as the intersection of different strategies represented by the variables cooperate-with-same? and cooperate-with-different?. This intersection in turn is represented in the form of conditionals. This is eloquently expressed in the section of the code in which each agent-type is given a particular shape, as shown in Figure 5.1.

In addition, there are other variables used in Figure 5.1, such as probability of reproduction (PTR), cost-of-giving, and gain-of-receiving. Collectively, the effect of these variables is to determine whether each agent can reproduce, which in turn is shaped by the total energy of each agent. The total energy at each step is calculated by subtracting the cost of sharing from the gain of receiving during exchange interactions with the neighbors. Whether or not any exchange will take place, in turn, depends on the types of the interacting agents, as defined by the strategies in Figure 5.3.

It is also important to note here that even the properties, behaviors, and interactions between agents are abstractions in a more basic, representational sense. For example, the color and shape of the agents are not designed to represent colors and shapes of people; they are representations of the different strategies (altruist, ethnocentric, cosmopolitan, and egoist). "Immigrants" in the simulation are not direct representations of human immigrants moving to a new country; they represent the appearance of strategies in a field of negotiation.

Similarly, reproduction in the simulation does not represent human reproduction; it represents the replication of successful strategies—a commonly used approach in evolutionary game theory. At a deeper level, the strategies themselves are abstractions that have been used to model different phenomena using different forms of modeling. Three examples of such phenomena are conflict and war,[34] and consumer choice,[35] and elections.[36]

```
to interact
  ask turtles-on neighbors4
  [
        if color = [color] of myself
        [
            if [cooperate-with-same?] of myself
            [
              ask myself [ set ptr ptr - cost-of-giving ]
              set ptr ptr + gain-of-receiving
            ]
        ]

        if color != [color] of myself
        [
          if [cooperate-with-different?] of myself
          [
                ask myself [ set ptr ptr - cost-of-giving ]
                set ptr ptr + gain-of-receiving
          ]
        ]
  ]
end
```

Figure 5.1
Wilensky's implementation of the ethnocentrism model

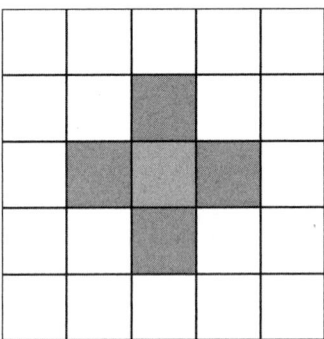

Figure 5.2
Von Neumann neighborhood of the square in the center is represented by the four colored blocks around it.

```
;; make sure the shape matches the strategy
to update-shape
  ;; if the agent cooperates with same they are a circle
  ifelse cooperate-with-same? [
    ifelse cooperate-with-different?
      [ set shape "circle" ]      ;; filled in circle (altruist)
      [ set shape "circle 2" ]  ;; empty circle (ethnocentric)
  ]
  ;; if the agent doesn't cooperate with same they are a square
  [
    ifelse cooperate-with-different?
      [ set shape "square" ]      ;; filled in square (cosmopolitan)
      [ set shape "square 2" ]  ;; empty square (egoist)
  ]
end
```

Figure 5.3
Wilensky's NetLogo implementation of strategies using different shapes of the agent

This duality of meanings of the words such as immigrants, reproduction, and even ethnocentrism, which are used both as computer code and in natural language, does not necessarily represent or lead to a crisis of understanding. Instead, it offers us rich possibilities for deepening and expanding the scope of learning and teaching with such simulations, as we discuss later in this chapter. It is at this mileu of artificial and natural languages that the experience of code and coding becomes powerful. This is also a transitional space, a leaky boundary between spoken language and computer code, which we believe can be an epistemological asset, not necessarily a liability. This is where pedagogical design comes in, so that the transparency and alterity of computational languages can be harnessed carefully and meaningfully through recontextualizing these abstractions.

5.3.2 Ethnocentrism as Emergence

A key characteristic of Hammond and Axelrod's model[37] is the persistent dominance of ethnocentric strategies over time, even as conditions in the simulation are altered. This is particularly striking given that at the early stages of the simulation, both the ethnocentric and the fully cooperating strategies are well represented in terms of agent numbers, but over time, ethnocentrism emerges as the dominant and stable strategy.

How does this happen? How does one particular strategy *emerge* over time as the dominant strategy, even when there is no initial bias toward it? In order to answer this question, we need to first understand the dynamics within the simulation, that is, the ways in which each agent in the simulation interacts with others. While the model runs, at every moment new agents (that

Hammond and Axelrod termed "immigrants") appear in random locations in the simulation. Each agent begins with a prespecified chance of reproducing, which is positively or negatively affected by the agent's choice to receive help from or give help to their neighboring agents. At every time step in the simulation, an agent is randomly selected to reproduce an offspring that inherits the same characteristics ("traits"). Whether or not the agent will actually reproduce, however, depends on the net probability of reproduction of that agent. Finally, each agent has a chance of dying, which makes room for future "offspring" and "immigrant" agents. So, let us now summarize this in terms of the possible actions that each agent in the simulation can undertake. At each time step, each agent can perform any one of the following actions: (1) reproduce and create a new agent with the same traits (an "offspring") in a neighboring patch, (2) remain static (do nothing), or (3) die.

As the simulation progresses, the dominance of particular strategies *emerges* through the interactions between different agents, as each individual agent follows the rules of interaction explained above. During the early stages in a simulation, new agents (immigrants) create regions of similar agents. Note that because any new agent cooperating with its own type is likely to produce a faster growing region than the one not cooperating, these homogeneous clusters tend to thrive. During these early stages, interactions *within* each region are more dominant in shaping the overall dynamics of the simulation, and each strategy seems to be represented fairly uniformly between these different regions. However, once the space in the simulated world is almost filled, the dynamics of the simulation are governed by the interactions between the adjacent regions with different attributes, where each region represents contiguous agents sharing the same color and strategy.

Hammond and Axelrod[38] found that the most important aspect of the regional dynamics in this simulation is the tendency of an ethnocentric region to expand at the expense of a differently populated region using any one of the other three strategies. This is an *emergent* outcome because it is a collective outcome that results from interactions between individual agents, even when no particular individual strategy is seeking to dominate other strategies. They found that after a transient period, ethnocentric agents dominate the population; humanitarians become the second most common agents; and both cosmopolitan and selfish agents become extremely uncommon. For example, when an ethnocentric agent interacts with a neighboring egoist, both defect. The ethnocentric agent, however, does better overall because it gets help from other ethnocentric agents in its region, but the egoist gets no help from other egoist agents in its region. As a result, over time, an ethnocentric region expands at the expense of an adjacent region of egoists.

   In the following section, we will examine the lives that computational models take on as other people—professional scientists as well as preservice teachers— reinterpret and revise these models. A common theme across the chapters in this book is that the experience of computational abstractions involves grounding computational abstractions in disciplinary contexts. The cases we present in the following section illustrate how such disciplinary grounding of computational abstractions can offer a potential pathway for engaging in critical conversations.

5.4   Grounding Computational Emergence Critically

The central point we make in this section is that recontextualization plays an important role in the *experience* of computational abstractions for modeling sociopolitical emergence, both in the professional practice of computational scientists as well as in pedagogical spaces. In section 5.4.1, we explain how this is evident in the work of computational scientists, and in section 5.4.2, we explain how the notion of transitional othering can offer similar opportunities in pedagogical contexts.

5.4.1   Abstractions as Recontextualization: A View from Computational Science

In order to understand how computational scientists *experience* computational abstractions, we begin with a view of the afterlife of Axelrod and colleagues' earlier work on ethnocentrism. In a recent paper, computational scientists (De and colleagues)[39] extended Hammond and Axelrod's model of ethnocentrism by recontextualizing the model to explain a rapid decline in violence and outgroup conflict over the past few centuries of human civilization. The disciplinary contexts in which De and colleagues situate their work brings together insights from recent statistical analysis of the historical data on human conflicts, as well as social and cultural psychology. They position Hammond and Axelrod's simulation as *group-entitative*, because the agents had perceivable group tags, interacted with their neighbors using Prisoner's Dilemma games, and could act differently toward in-group and out-group members. They investigated whether they could extend Hammond and Axelrod's simulation so that the computational agents could base their actions on knowledge of other individuals rather than group tags (*individual-entitative*), and whether such a strategy would be favorable evolutionarily.

But before we proceed further, it is important to recognize that recontextualization played an important role even in Hammond and Axelrod's original work. It is noteworthy that Hammond and Axelrod themselves made it clear that their simulation was "not intended as a realistic portrayal of specific social behaviors."[40] Instead, they used an evolutionary framework that had been used by other researchers in political science and computer science to study adaptation.[41] However, even such an abstraction is not devoid of disciplinary grounding. As the authors explained next, their use of evolutionary models was in turn based on research in psychology and anthropology that argues that "broad and potentially heritable universals of human thinking" result from evolution, which in turn leaves the human mind "predisposed to react in certain ways."[42] The authors cited anthropological research that specifically argued that in-group bias is one such "innate universal predisposition," thereby positing that nationalisms and racisms are likely "hypertrophies" of such in-group bias in the form of ethnocentrism.[43] And finally, it is very important to remember that even this model was an adaptation of an earlier computational model that the same authors previously developed to study the evolution of altruism in biological settings. In this earlier model, similarities between individuals were interpreted in terms of considerations observable even by microorganisms, such as cell surface structure or pheromones, whereas in the ethnocentrism simulation, observable characteristics that people may find socially relevant (such as skin color or language) were the basis of judgments about similarity.

Therefore, even though one might think about Hammond and Axelrod's model in terms of the computational abstractions such as the algorithms (e.g., Prisoner's Dilemma) and data structures (e.g., group tags) used to simulate the ethnocentric behaviors, these abstractions only become meaningful through disciplinary grounding and recontextualization. This is particularly poignant given that their approach is of universalist nature (e.g., potentially attempting to explain racism in terms of "innate universal predisposition" of in-group favoritism), a general approach which has been rightfully critiqued for its omission of violence against and colonization of people of color.[44] Our point here is that even a universalist computational approach is rendered meaningful through disciplinary forms of recontextualization (e.g., using anthropological research to justify this approach).

We can now understand De and colleagues' work as yet another instance of disciplinary recontextualization of Hammond and Axelrod's work. They directly call into question what Hammond and Axelrod's model represents in terms of social behaviors and propose another recontextualization based on scholarship in social and cultural psychology, especially in contexts in which the agents are highly mobile. They grounded their model in research in cultural psychology that shows that in high-mobility contexts, individuals change relationships often, building new relationships and severing unwanted relationships quite malleably.[45] This is also a progressively richer form of recontextualization (compared to Hammond and Axelrod's work), as it brings into account a greater heterogeneity of human experience.

Building on this work, De and colleagues argued that in societies where people are highly mobile (that is, they move frequently), having a broad network of weak ties and being open toward strangers is highly adaptive. Using evolutionary game-theoretic models similar to Hammond and Axelrod, but recontextualizing these models in a different body of scholarship in social sciences, they showed that out-group hostility is dramatically reduced by mobility, thus challenging the "inevitablity" of ethnocentrism that was a central outcome of Hammond and Axelrod's work. A key finding of their model is that as mobility increases, the evolutionary pressures shift to favor individual-entitative agents. In other words, individuals are more likely to adopt strategies for assessing trustworthiness of others in highly mobile contexts, for which they rely on their memories of previous interactions with other individuals, rather than group tags. In contrast, in low-mobility contexts (e.g., in contexts similar to Hammond and Axelrod's simulation of ethnocentrism), individuals have far fewer opportunities to form new relationships. As a result, severing existing relationships can have adverse effects such as being ostracized from one's social circle. Therefore, in low-mobility contexts, individuals may rely predominantly on group tags to examine possibilities of cooperation with others and thus act in a group-entitative manner.

So, even though both the papers address issues of cooperation and conflict in human societies through computational modeling, each paper uses different ways of recontextualizing computational abstractions in disciplinary theories and perspectives, including empirical studies in the social sciences, in order to represent these issues. These forms of recontextualization, in turn, have significant impacts on what forms of abstractions can be used, for example, as represented in the form of different algorithmic rules and data representations (e.g., mobility and individual-entitativity) used by De and colleagues compared to the original Hammond and Axelrod model. To isolate the computational abstractions from the phenomenological contexts in which they become meaningful is simply an inappropriate reduction of the modeling process and experience. Furthermore, as we shall see in the following sections, an explicit focus on recontextualization can be pedagogically important.

5.4.2   Recontextualization as Transitional Othering

To grasp the significance of this *transitional othering*, we must take a moment and think carefully about the phenomenology, that is, our *sense experience* of talking about race, both in the K–12 classroom and otherwise. The overarching cultural and systemic pushes to silence conversations about race in the curriculum results in what Pollock termed *colormute* behaviors.[46] It is therefore no surprise that this is also the way in which computing has largely come to be used in the classroom. If we are to avoid colormute behavior, race and race talk must be heard and discussed as part of the computational modeling activities. In fact, sociologists have noted that even in everyday encounters, uncertainty in dealing with perceived stereotypical (including racial) differences is a quite common experience.[47] Positioning both the "self" and the "other" in these moments is an experience that is usually rife with uncertainties, and as Hess and McAvoy[48] have pointed out, this often results in the avoidance of political issues in classroom conversations.

So how do we bridge this gap through transitional othering? It is a critical phenomenological reframing of Papert's vision of computational agents as transitional objects.[49] That is, virtual agents such as the Logo Turtle act as mediational objects, which learners can use as proximal projections of their lived experiences. Drawing a circle with the Logo Turtle occurs in a body-syntonic fashion: just like the child, the Turtle moves a little, turns a little, and draws a circle with its body. The Logo Turtle, Papert argued, is thus transitional in the sense that it lives both on the computer and with the child. But it is also in another sense that the Turtle can be transitional. When computational agents are used to simulate or represent socioscientific phenomena such as ethnocentrism and racial distribution of the US population, computational agents serve both as representations of the self and the other.[50] The other is not merely someone else. As the critical scholar Patricia Hill-Collins has argued, it is an experience of being different from or alien to the identity (including social identities) of the self.[51]

To summarize what we have said so far: the transitional nature of computational agents can provide students and teachers with a space where they can explore complex critical sociopolitical issues without directly situating the phenomena in their own personal lives and yet, at the same time, use at least some of their lived experiences to make sense of the simulated interactions. At its core, transitional othering involves perspectival thinking by allowing us to bring in our own perspectives, and also by creating a space for us to take on the perspectives of other people (represented by "other" agents) whom we would normally not identify with ourselves in our daily experiences. Furthermore, it is not necessary that we take on these perspectives all by ourselves; the sharing of different perspectives can also happen in the form of group discussions, similar to what we encountered in chapter 3. That is, the lives of "others" can also become explicit as different students share their different interpretations of the stories underlying the actions, possible histories, and interactions of the agents. This is a liminal space, which is at once deeply personal and also shared in a safe space in the form of small group discussions, as we describe later in this chapter.

In this chapter, we present a learning environment and a pedagogical approach to support such experiences. We first reported this environment and an earlier version of our analysis elsewhere.[52] Here we present a significant reanalysis of the data, our goal being to draw attention to the fundamental role that recontextualization plays in the experience of computational abstractions and how this can be leveraged for engaging critically with computing.

5.5 Designing for Transitional Othering

5.5.1 Setting and Tools

We now provide an example of the learning environment designed to support *transitional othering*. The setting of this study was a semester-long course in human geography in the Secondary Education Social Studies program in a major private university in the Mid-South. Nineteen students (eight senior undergraduate students and 11 master's students), all preservice teachers, attended the course. This study took place in the middle of the course. In class sessions prior to the study, students had discussed the relevance of zoning and the impact of the Civil War and Jim Crow laws on urban segregation in the Southern US cities. Our work was motivated by the fact that despite such curricular experiences, prior to introducing multiagent representations, when we asked the class to design lesson plans for teaching about the Civil War and political zoning in the South, none of the students explicitly mentioned or designed any activity that involved explicitly reasoning about race and racial inequalities. Our goal was to see if and how students' experiences with agent-based computational representations of segregation and ethnocentrism could reshape their lesson plan designs, with the hope that such interactions would facilitate a more direct engagement with issues related to race and racism, as well as other forms of inequality.

We adopted Wilensky and Rand's implementation[53] of Axelrod and Hammond's model of ethnocentrism[54] in the NetLogo multiagent platform.[55] We developed a set of programming blocks in the ViMAP block-based programming and modeling platform.[56] Using the ViMAP programming blocks, students could control the NetLogo code and the simulation, and doing so also made the underlying rules explicit to them (Figure 5.4). Overall results from running the simulation were displayed in the form of multiple graphs in separate windows. These graphs showed how the populations of the different strategies would compare with each other, as the simulation unfolded.

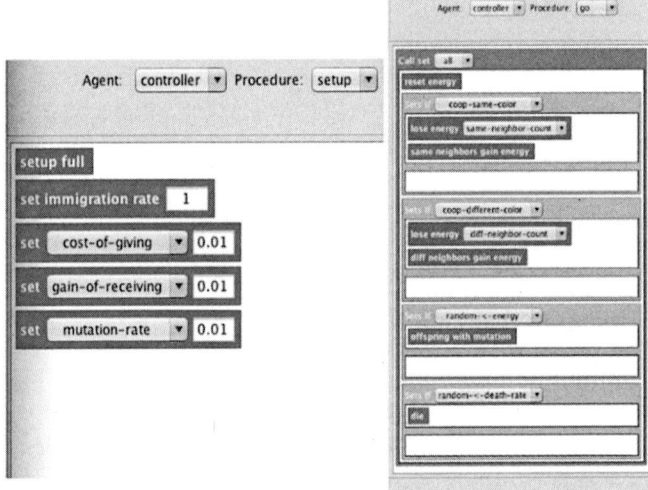

Figure 5.4
A ViMAP implementation of Wilensky's code

Figure 5.5
The Racial Dot Map: A screenshot of racial distribution in Chicago

In order to provide supports for recontextualization of the computational model and abstractions, we used the Racial Dot Map[57] shown in Figure 5.5. This map provides an easily accessible online visualization of the geographic distribution, population density, and racial diversity in every neighborhood in the US. The map displays over 300 million dots (equal to the population of the US as reported in the 2010 US census), using one dot in the form of a circle for each person at their residential address provided on the census. The stated ethnicity of each person on the census is represented by the color of each circle.

In this map, each dot can be conceptualized as an anonymous agent, similar to the agents in the ViMAP/NetLogo simulation based on Hammond and Axelrod's model. In addition, the map allows users to zoom in and visualize the racial density and distribution at the level of each city and neighborhood. For example, zooming in on the city of Chicago reveals that the neighborhoods are fairly diverse around downtown, and become heavily segregated farther away from the city center.

We believed that reasoning about anonymous agents in the Racial Dot Map would enable the students to bring in their relevant *lived-in* experiences as part of the conversation, because they would be able to identify their lived-in areas (cities and/or neighborhoods) in the map. The use of the ViMAP simulation then served a complementary role, because students could then "see" the emergent behaviors and the underlying agent-level interactions as representations of possible underlying mechanisms of real-world residential segregation. Anonymous computational agents—dots in the Racial Dot Map and ViMAP agents in the simulation—served as transitional objects in both the tools. This liminality is epistemologically significant, and understanding how this liminality can be of pedagogical use can in turn offer us a richer understanding of what computational abstractions can mean, as well as how computational modeling can help us engage in difficult conversations about racial and other forms of intersectional inequalities.

5.5.2   Analysis

In the following subsection, we present brief discussions of three forms of discourse that we observed in the class. As described elsewhere, the data for this project was collected in the form of students' written work (activity guides and their premodeling and postmodeling lesson plans) as well as focus group in-depth interviews conducted with them as they were interacting with the simulations and the Racial Dot Map and redesigning their lesson plans.

The analysis we present here is focused primarily on illustrating the students' *sense experiences* of the ViMAP simulation and the Racial Dot Map as tools to think about urban segregation. The comparison of their pre- and postmodeling lesson plans is rather obvious, in that we expected to see significant growth in terms of their ability to bring in issues of race and marginalization in their proposed activities around the Civil War and zoning in the Southern US. This shift is certainly what we saw, even to a high degree of statistical significance, as we have already reported elsewhere.[58] Here, our focus is more on highlighting the *experience* of transitional othering and recontextualization *while* students were interacting with the simulation and the Racial Dot Map, which we believe to be at the center of this shift.

5.6   The Experience of Recontextualization and Transitional Othering

5.6.1   Early Noticings: The Dominance of Ethnocentrism

During the class, Pratim introduced the students to the ViMAP simulation through an interactive presentation that lasted roughly an hour. The simulation was positioned as a tool for provoking thoughts, and the representativeness of the agents was specifically characterized as open-ended. Pratim encouraged the students to discuss with their neighbors what the agents should stand in for. Students sat in small groups of three or four.

An advantage of block-based programming and, in particular, the ViMAP platform is that the programming commands involve both domain-general and domain-specific primitives. Instead of providing students with a general programming language (e.g., C or C++ or even Scratch), the ViMAP ethnocentrism simulation was specifically designed so that the library of programming blocks contained only the blocks that were customized for the simulation. Students could select blocks from the library following the class discussion and activity guide, and the role of the teacher (in this case, Pratim) was to engage the students in meaningful and illustrative class discussions to think carefully about the meaning of relevant programming commands and about the relationship between the emergent outcomes in the simulation and the programming commands.

This means that the conditional rules represented in Figure 5.1 were immediately accessible to the students through the ViMAP command blocks. The block-based nature of the programming also reduced chances of syntactic errors, which is particularly important for classrooms in which students have no prior experience with programming, and are also limited in time. Each student was provided a link to the downloadable ViMAP simulation, and they worked along with Pratim to create the program as shown in Figure 5.4. Pratim introduced each programming block through a "reflective toss," that is, by explaining how the programming block affects the behavior of agents in the simulation, and then asking the students to change the relevant parameters and predict how that might affect the behavior of the agent (e.g., changing the cost-of-giving to a much higher number might make agents more ethnocentric). Once the entire program was constructed, Pratim asked students to run the simulation on their own computers under a variety of conditions, notice the graphs that tracked the population of each type of agent, and discuss with their neighbors possible explanations of the outcomes. Students were also given activity guides in which each ViMAP programming command was explained and there were prompts for noting their explanations of how the changes to the ViMAP code affected the outcomes of their models.

As students ran their simulations with the initial set of parameters specified by Pratim, they started talking about the dominance of the ethnocentric strategy. For the same set of parameters, running the ViMAP simulation a few times would yield different sets of outcomes, because of the different initial locations of the agents in each run of the simulation. Some students also tried to run the simulation a few different times with different parameters each time. In both cases, students overwhelmingly reported that the ethnocentric strategies dominated in their simulations.

An example of a student's written work is shown in Figure 5.6. This example illustrates that the student was engaged in altering different parameters in the simulation and observing the effects of the changes in their simulation. However, we also noticed that across the class, all the students were focused on observing the aggregate-level outcomes (population graphs, indicated by the "STRATEGY-COUNTS" plot in the activity guide). We did find that in reporting which type of agents had the highest population during different runs of the simulation, several students had referred to the agents as "people" but they did not identify individual agents. We do not consider this to be evidence of transitional othering; this is simply a superficial mention of representativeness of computational agents, and does not do any pedagogical work for the students.

Only two students went a step further in exploring the human-agent analogy. In reporting which strategy won, one student wrote, "Racists win under most conditions." The same student also noted that the simulation was taking place in a "closed space." That is, in the simulation, there was a finite number of locations that could be occupied by the computational agents, and this was similar to the real world's having finite resources. He then raised the question of how "empathy" and "emotions" in agents might affect their interactions and overall survival, because he had read somewhere that empathetic people have a higher chance of survival in a world with finite resources. Another student also used the language of "caring" to explain why ethnocentric agents typically

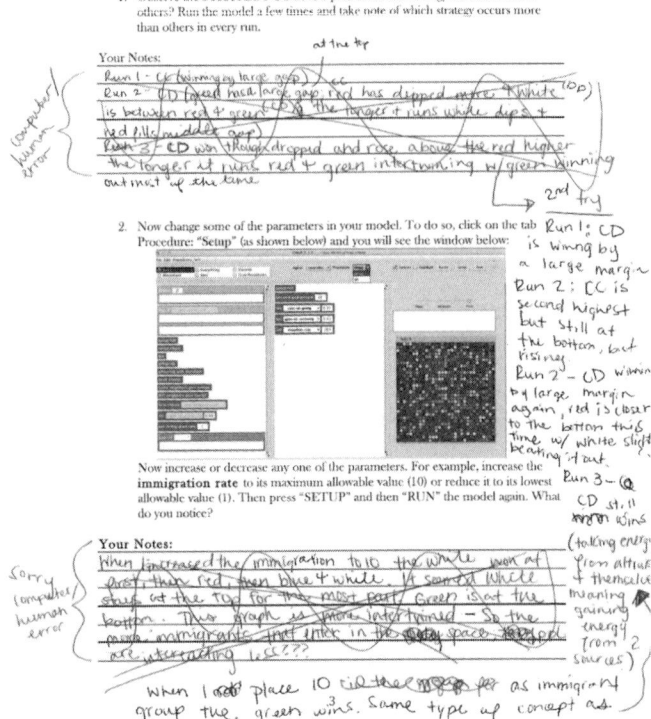

Figure 5.6
Example of student work with ViMAP simulations

won. He explained that this happens because "people tend to care for similar people, because they see themselves in others who are similar."

Near the end of the hour, students were introduced to the Racial Dot Map as another tool that they could refer to, manipulate, and use, in addition to the ViMAP simulation. Their goal for the remainder of the class was to redesign their lesson plans on the Civil War and zoning.

5.6.2   Recontextualized Noticings: Clustering, Contrasts, the Self, and the Other

Once introduced to the Racial Dot Map, students began running and comparing their work on the ViMAP simulation with zoomed-in views of the Racial Dot Map. This comparison served the important role of recontextualization, similarly to what we saw in the work of the computational social scientists. The conversations were both about the computational rules of interaction between agents in the ViMAP simulation, the emergent outcomes, and their noticings of patterns of urban segregation in cities across the US.

The following figures present a side-by-side comparison of two images: a screenshot of the ViMAP simulation showing several ethnocentric clusters (agents shaped as straight lines are ethnocentric in Figure 5.8), and a zoomed-in view of the vicinity of downtown Minneapolis in the Racial Dot Map (Figure 5.7). One of the groups noticed that the areas around downtown Minneapolis and downtown Nashville (especially around the hospitals and the major universities) were racially diverse, whereas the ViMAP simulation almost always eventually resulted in predominantly ethnocentric clusters of agents scattered throughout the simulation.

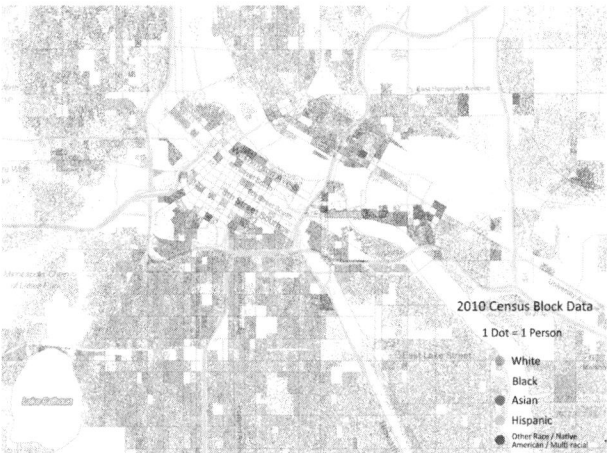

Figure 5.7
Minneapolis in Racial Dot Map

Noticing this contrast led to conversations that brought to fore two forms of reasoning: (1) conversations about the computational rules that the agents were following in the simulation that would lead to such clustering, and (2) attempts to reconcile these noticings in terms of the students' own lived experiences and perceptions of inequality and racism. It is also important to note that these two forms of reasoning were intertwined with each other.

For example,[59] several participants discussed that urban segregation was common in cities in both the Northern and Southern US, although a general trend was that places near the city center were more diverse. However, in the simulation, clusters of similar agents were spread throughout. One of the participants (Tami) noted that perhaps one way to explain this disparity is that in

Figure 5.8
A sample screenshot showing clusters of ethnocentric agents (straight lines)

the real world (unlike in the simulation), the same people may exhibit both ethnocentric and cosmopolitan forms of behavior. Tami was also concerned by the persistent dominance of ethnocentric agents in the simulation, because it did not resonate with her own experience of how in some cases the socio-economic and racial disparities may decrease over time. Tami noted that this was true in her neighborhood where she grew up, which became more diverse over time. Tami also noted that this was not a general trend but that she wanted the simulation to better represent both possibilities. Tami's concerns involved rethinking the computational rules obeyed by the agents in the simulation and making them more dynamic (that is, contextually determined), in a manner similar to that which De and colleagues suggested.

In designing activities for their future students, all participants noted that us-ing the Racial Dot Map could help situate students' own personal experiences as relevant to the modeling and learning experiences. As we have pointed out elsewhere,[60] this involved conceptualizing the computational agents as a rep-resentation of both the "self" and the "other." For example, one participant (Manny) noted that:

> *Manny:* Also, you want to give them [future students] the map as a personal experience, but give them an out-of-body experience, like, for me being a student, I would just be a blue dot. And I don't want to think of my life as just a blue dot or be an empty square, to give them this, like, out-of-body experience to look down on themselves. Like, I'm a blue dot, but it's also important to think about the green dots, the red dots, the yellow dots, the brown dots, or, like, think about how empty circles interact, empty squares,

so, like, seeing them from a different perspective and acknowledging and challenging themselves, I think from not the normal perspective. So yeah, like find your house, but then, like, it's not just about your house.

Here, Manny is referring to the modeling experience as both a "personal" and an "out-of-body" experience. Manny is referring to agents both in the Racial Dot Map (". . . a blue dot . . .") and the ViMAP simulation ("or, like, think about how empty circles interact, empty squares . . ."). A key characteristic of Manny's explanation here is his noticing of the importance of being able to "see" oneself in the Map and the simulation and also think about potential interactions with others. Also notable here is that for Manny and his group, and for most other groups, these conversations began to take place only after they started interacting with the Racial Dot Map. The agents in the ViMAP simulations are located on a hypothetical computational space, whereas each dot on the Racial Dot Map represents an American resident at the location of their home on the US map. The latter makes it difficult not to place or find oneself on the map, whereas the former lends itself to conversations similar to the conversation discussed in the previous paragraph.

It is the juxtaposition of these two forms of representation that was helpful in supporting deep engagement about lived experiences of the participants themselves, as well as the imagined histories and interactions of the computational agents (both the dots and the ViMAP agents). In contrast, as noted in the previous section, focusing only on the agents and the aggregate behaviors of the simulation does not necessarily lead to such forms of engagement. The persistent dominance of the ethnocentric strategy in the simulation became more meaningful as the participants began noticing contrasts and similarities in the Racial Dot Map. The juxtaposition created opportunities to both "dive in" and "step out"—epistemic moves that are critical for developing understandings of complexity.[61] In the following section, we illustrate how reasoning about both a historical and a contemporary issue further shaped this experience of transitional othering, and supported the teacher candidates in developing an understanding of computational rules and variables as somewhat generalizable.

5.6.3 Recontextualizing as Modeling the Past and the Future

Students also noted that leveraging thinking in terms of computational abstractions (variables) such as COST-OF-GIVING and GAIN-OF-RECEIVING as representations of economic exchanges could be a productive way forward for teachers and students to think deeply about economic inequality, even from a historical perspective. For example, one participant (Jamie) noted that thinking about the implications that these variables have on the interactions between agents in the simulation could help students engage in conversations about how economic incentives and ideologies may have sustained slavery in the Southern US. She said that using the simulation, students could explore how the "*ideology of economic incentives can produce a give and take relationship, in which one group may exploit another.*" Jamie's group further argued that using the simulation could therefore lead to reframing discussions about the Civil War to focus on the economic underpinnings of slavery, given the centrality of intergroup exploitation based on economic motives. Jamie also noted that intergroup exploitation and the ideology of economic incentives could also be a productive way for students to think about "*labor unions and, like, labor movements in the 20s, 30s, 40s.*" This implies that Jamie and her group were beginning to identify intergroup economic exploitation as a generalizable mechanism that could be used to support student thinking about the underlying issues behind the Civil War and labor union movements in the US.[62]

A particularly striking phenomenon in these conversations was the temporal juxtaposition of conversations about historical events and the participants' own lived experiences. Undoubtedly, the activity design was an important factor in these conversations' unfolding one after another. However, we believe that the generalizabilty of the emergent mechanisms was also aided by the participants' being able to "see" the personal relevance, particularly in the context of noticing racial segregation in the Racial Dot Map. For example, the discussion on the Civil War in Jamie's group was preceded and succeeded by discussions on how residential maps around industrial areas tended to be monolithic in terms of race, and how the histories of race and class in the US are deeply intertwined. The following is another illustrative example, in which we discuss Manny's group (Manny, Robert, and Tandy) engaged in reasoning both about their lived experiences and the historical dimension:

> *Robert:* I was just more thinking of, you want to live with people who have a similar plight as you. Or just a similar, you grow up in a certain neighborhood. Like, I grew up in south Nashville and I will eventually move back there. So, I think that the simulation helps you understand, like, this might be a dark way of looking at it, but at our most basic like level, we're concerned about our own survival, so you see a map of people of a similar view, you might see yourself in them unless you want to help them. So, a certain neighborhood will have a certain kind of people because they want to see themselves so they want that neighborhood to do better because they live there and they connect with the people there.
>
> *Tandy:* I think, so we haven't had time to really discuss what we're going to do with the Civil War yet but that we're going to look at the geography . . . and how the geography in the South, even if it influences the history of the South . . . in the last would be a good idea so we can demonstrate how ethnocentric views really end up being most prevalent in a given space and therefore it will be easier to understand why slavery wasn't necessarily

considered an evil thing in the South right before the Civil War and how so many people . . .

Pratim: Can you talk a little more about that? That's very interesting.

Tandy: Yeah, so, if you have a group of people in an area who believe one certain thing, and they're not going to let that view be influenced by other people. I don't know. It's self-reinforcing.

In this excerpt, Robert responds to Pratim's question to the group about how the map and the simulation are helping them think about lessons on both zoning and Civil War. He begins by revealing something about his personal life: he grew up in south Nashville, and *"will eventually move back there."* Robert then explains that the simulation helps us understand *"how a certain neighborhood will have a certain kind of people because they want to see themselves so they want that neighborhood to do better because they live there and they connect with the people there . . ."* It is important to note that Robert's statement that he might move back to the neighborhood he grew up in is indicative of the influence that life course experiences have in neighborhood selection during adulthood in the US. Sociologists and economists have empirically demonstrated that limited experience with integrated neighborhoods during adolescence among non-Whites and limited geographical mobility among all young adults help to perpetuate segregation.[63] So, Robert's noticing here could also be seen as an opportunity for diving even deeper into issues of segregation from the perspective that residential segregation and neighborhood qualities are, to a significant extent, perpetuated over the early stages of the life course.[64]

Tandy followed Robert with an explanation of how the map and the simulation could be used in teaching about the Civil War and slavery, which also highlights how their noticings of clustering play an important role. He emphasized the geographical (spatial) dimension of both the simulation and the Racial Dot Map. He noted, *"we can demonstrate how ethnocentric views really end up being most prevalent in a given space and therefore it will be easier to understand why slavery wasn't necessarily considered an evil thing in the South right before the Civil War and how so many people. . ."* What stands out here is Tandy's realization that the simulation might help students understand why slavery wasn't "necessarily considered an evil thing," despite its being so prevalent. The spatial density of similar strategies in the simulation, according to Tandy, can help us understand the self-reinforcing nature of ethnocentrism: *"if you have a group of people in an area who believe one certain thing, and they're not going to that let view be influenced by other people. I don't know. It's self-reinforcing."*

As the conversation continues, Robert and Manny point out how noticing clusters of agents (dots) could play an important role in engaging in conversations about racial segregation in the US:

> *Robert:* I think these were all tools to compare regions of the country. Like, looking at this map and the program, South has always been a teaching history—the Civil War, Jim Crow laws, and Civil Rights Movement, but then you look at the map like this, most of the diversity in the US, or for the larger portion of it, is in the South, and so it's like we're all separated within the city or whatever but there is still a lot more interaction with people, instead of somewhere like Minnesota or something. . .
>
> *Manny:* It's like you were redefining cooperation. Like, in the model, cooperation is the exchange of ideas, and so therefore, ethnocentrism would be only receiving, no only giving your ideas to your same race. And receiving other ideas from other people but like that would perpetuate, only giving idea, only giving or sharing with the same race, would self perpetuate. I kind of talked myself in a circle there.
>
> *Robert:* Like if everyone around you thinks something, you're going to think the same thing and that's how societies are societies. . .

Robert brings up the idea that both the simulation and the map were "*tools to compare regions of the country.*" He argues that while the Southern states have always been positioned in history (classrooms) in terms of Civil War, civil rights movement, and Jim Crow laws, he noticed that urban centers in several states in the Southern US are less segregated than Northern states such as Minnesota. Robert's observation that most of the diversity of the US is in the South could easily be debated for its exactness. However, it is indeed true that the Southern states are progressively becoming more diverse, and according to the US census data, both Memphis, Tennessee, and Houston, Texas, are among the top 10 most diverse metropolitan areas in the US.[65]

The implicit push in Robert's explanation here is quite profound: by looking carefully at the geographical picture of ethnic segregation across the US, one could go beyond superficial geographical stereotypes (e.g., the South is more segregated than the North) and instead engage in a deeper conversation about the potential reasons behind ethnic segregation. This becomes evident in Manny's explanation, who continues Robert's line of reasoning and notes how the computational rules in the simulation were being recontextualized in their explanation. He noted that while cooperation was defined in the simulation as the exchange of ideas, in their conversation, ethnocentrism could be understood only as "*receiving*" ideas from other people in "*your same race,*" thereby leading to the self-perpetuating nature of ethnocentrism. This is further clarified by Robert, who noted that it is the similarity among people who are around us that creates societies.

Overall, it becomes clear in this interaction that Robert, Manny, and Tandy are reinterpreting the computational rules of interaction in Hammond and Axelrod's original simulation through recontextualizing them in two different situations, one that is historical and the other closer to their lived experiences. In doing so, they are also developing a sense of how these rules are generalizable representations that can extend beyond the immediate context of the simulation. The dots on the Racial Dot Map and the agents in the simulation are at once their projected selves, as well as representations of others. The computational agents are liminal in this sense, allowing Robert to share his personal history with a particular Nashville neighborhood and also to be able to notice how clusterings of people of the same ethnicities compare across different metropolitan areas. It was through these conversations that the group was able to articulate how the variables in the simulation would need to be reinterpreted for different contexts. We believe that this form of progressive recontextualization makes computational rules *generalizable* as part of our learning experiences, grounded in critical perspectives.

5.7 Discussion

5.7.1 Projectivity and Transitional Othering

Critical scholars in the learning sciences position educational design as a journey toward reimagining futures through future-making[66] that involves actively working toward social and moral justice. This in turn involves both imagining a particular future and consciously trying to contribute to it, through creating new spaces and opportunities for critical discourse that challenges social and political systems of hegemonic oppression. This chapter highlights how the experience of projection, as reflected in our illustrations of *transitional othering*, can offer such opportunities, which at once enable future teachers to position themselves in complex simulations of social inequalities and to challenge what is "given" through *historically grounded critical discourse* with the simulations. The *sphere of givenness* that is being challenged here includes both the culture of colormute behavior and pedagogy in preservice teacher classrooms and the culture of politically neutral and colormute use of computational simulations in K–12 contexts (including K–12 teacher education). In this sense, the experience of projection is a critical phenomenological endeavor.

Projection is fundamental to human experience, including our experiences of race. Dewey[67] characterized *projecting* into the future as a certain form of imagination that is a fundamental and necessary form of experience. He wrote

> Experience in its vital form is experimental, an effort to change the given; it is characterized by *projection*, by reaching forward into the unknown; connection with the future is its salient trait.

Race scholars have also noted the importance of projection and projectivity in shaping our experiences of race, particularly for racialized people (people of color). For example, Emirbrayer and Desmond noted that phenomenological philosophers position "social actors as 'thrown' into historically evolving situations and as projecting themselves into their own possibilities of being."[68] They argue that in such situations, projectivity can play an important role for racialized actors:

> Projectivity entails the imaginative generation by racial actors of possible alternate trajectories of action, in which received structures of thought, feeling, perception and action are creatively reconfigured in relation to actors' hopes, fears and desires for their futures. (p. 147)

In this view, projectivity allows racialized people to distance themselves from problematic habits and traditions and to challenge and reformulate those established schemas. Our work shows how multiagent simulations and computational representations can offer such opportunities by harnessing projectivity. Nearly a century back, George Herbert Mead termed this form of experience "distance experience," in which people move "beyond themselves" into the future, constructing imagined worlds for escaping the circularity of social reproduction of racial oppression.[69] Mead especially stressed the intersubjective aspect of projectivity, highlighting our ability to project our*selves* into the experience of others. Multiagent simulations and representations, such as the ones used in this chapter, provide dynamic and experientially rich opportunities for engaging in such forms of experience through transitional othering.

The cases we have discussed illustrate that the experience of *transitional othering* involved projection, as well as an effort to change the given. Along the first dimension, the participants' interpretations of the ViMAP code and the different forms of computational visualizations involved transitional othering, and projection of the self and the other constitute key elements of this experience. Far from being an experience of certitude, this is a liminal space. We have seen cases where they are at once identifying themselves as computational representations and, at the same time, attempting to see beyond their own experiences through adopting others' perspectives. Along the second dimension, we have seen how code was reinterpreted and recontextualized as participants in our study and professional researchers used similar computational representations as they attempted to imagine better futures. This is evident both in professional practice in computational social science and in the context of preservice teachers' attempts to design activities that would encourage conversations about race and inequality. De and colleagues' extension and reconceptualization of Hammond and Axelrod's work led to models of worlds in which ethnocentrism can be countered through greater mobility, and preservice teachers such as Tami and Manny saw multiagent-based representations and simulations of ethnocentrism as contexts and scaffolds for helping their future students develop a better understanding of critical issues such as racial and economic inequalities.

5.7.2 Transitional Othering as Critical Discourse

The relationship between transitional othering as a form of projectivity and as a form of critical discourse is also premised on the inherent ambiguity and transparency of computing and code as discourse. To remind the reader, Bakhtin noted that in ambiguous discourse, several meanings of the same utterance are to be taken on exactly the same level, and syntactic, semantic, and pragmatic ambiguity all lie within the scope of the experience of language. In transparent discourse, there is no attention given to the literal meaning. The connection with coding here is obvious, given that every computational representation *stands in* for something else. How are these forms of discourse evident in our analysis? We explain this next.

The syntactic ambiguity was easily overcome due to the block-based nature of the programming language, and this is important because previous research has shown that this form of ambiguity makes integrating coding in classroom teaching quite challenging. However, it is the pragmatic and semantic ambiguity that became important in the learning experiences of the preservice teachers. The pragmatic challenge of how to talk about race, as race scholars have argued, is commonly experienced in our everyday lives. In the classroom, as social studies educators have pointed out, discussing race and inequality is also a pragmatic challenge as much as it is an epistemological and pedagogical issue, given limited curricular support and scope for such issues. In addition, the growing racial and cultural disconnect between the teachers and students, as noted by social studies scholars, complicates the issue further. Our study shows that the transparency of computational representations, when taken up in discourse and supported through appropriate means for recontextualization, can be helpful in addressing the issue of pragmatic ambiguity of race talk and conversations about social inequalities.

The ViMAP agents and the Racial Dots take on multiple, imagined lives in the preservice teachers' discourse. They stand in for the participants themselves, their future students, their neighbors, and imagined others. Transparent discourse can also be ambiguous when it stands in for more than one referent, and we believe that this is certainly the case, for example, when Manny refers to the computational agents as both the "self" and the "other." So, this is a form of transparency and ambiguity that can be pedagogically helpful. Similarly, variables in the ViMAP simulation such as COST-OF-GIVING, GAIN-OF-RECEIVING, and so on, stood in for economic exchanges between people as well as a ground on which Jamie and her group began reframing conversations about the Civil War to focus on possible economic underpinnings of slavery.

The computational representations and abstractions used in the simulation, recontextualized through the use of the Racial Dot Map as well the pedagogical framing of the activity, thus became sites for productive transparent and ambiguous discourse, which in turn pushed forward the critical conversations. Papert's notion of the Turtle as the transitional object shines through our work, in which the ViMAP agents and the dots of different colors in the Racial Dot Map play the roles of transitional objects. However, it is in the experience of these transitional objects—*transitional othering*—in which they come to life. In this experience, the dots and the agents are both ourselves and others, and it is necessary and pedagogically powerful to take on both perspectives. Heterogeneity here is perspectival, as well as evident in the form of the different lived experiences across the classroom that the different participants brought into play in their discussions.

It is important to note that as Emirbrayer and Desmond argued,[70] projectivity is not the end but the beginning of the journey toward developing agency in contexts that involve racialization. Simply put, the experiences of transitional othering of the preservice teachers in our study, although powerful, represent only the beginning of this journey. Similarly to the research program developed by Lehrer and Schauble, we believe that integrating such experiences across the curriculum, over multiple years of the K–12 experience (and for preservice teachers, throughout the duration of their studies), is essential.

While we must work toward such futures, it is worthwhile revisiting the issue that motivated our work: racial and social inequalities are generally excluded from educational computing in K–12 classrooms and preservice teacher preparation, and there is a culture of silence around the deeply horrific history of racial violence and oppression in the US.[71] The specific context of our study was not different, as it grew out of our concern that an almost exclusively White group of preservice teachers was generally unwilling to break this unwritten code of silence, even in a course that focused on critical geography. In such settings, we posit that transitional othering could in fact provide opportunities of "distance experience" in which both students and teachers can feel comfortable to talk about race in terms of hypothesized or partially imagined experiences of the computational agents or the dots. And although this chapter illustrates how such experiences of code can be voiced and the central role of heterogeneity, it is also important to note that as mentioned in section 5.2, there are many omissions in working with theories and models of ethnocentrism. Furthermore, as we have discussed elsewhere,[72] a certain amount of presentism[73] is also inherent in attempting to model or represent historical events or phenomena. So, our goal is not to offer our work here as a settled computational representation of race or sociopolitical inequality—but as an open invitation to the field to continue such inquiry.

Notes

1. T. M. Philip, A. Gupta, A. Elby, & C. Turpen (2018). Why ideology matters for learning: A case of ideological convergence in an engineering ethics classroom discussion on drone warfare. *Journal of the Learning Sciences*, 27(2), 183–223.

2. M. A. Takeuchi, P. Sengupta, M. C. Shanahan, J. D. Adams, & M. Hachem (2020). Transdisciplinarity in STEM education: A critical review. *Studies in Science Education*, 56(2), 213–253.

3. T. M. Philip, M. C. Olivares-Pasillas, & J. Rocha (2016). Becoming racially literate about data and data-literate about race: Data visualizations in the classroom as a site of racial-ideological micro-contestations. *Cognition and Instruction*, 34(4), 361–388.

4. N. Allum, P. Sturgis, D. Tabourazi, & I. Brunton-Smith (2008). Science knowledge and attitudes across cultures: A meta-analysis. *Public Understanding of Science*, 17(1), 35–54.

5. Philip et al, Becoming racially literate, 2016.

6. S. U. Noble (2018). *Algorithms of oppression: How search engines reinforce racism*. NYU Press.

7. K. Truong (2020). Red Hat audit to 'eradicate' problematic language in its code *Vice*. Available at: https://www.vice.com/en/article/qj4jqb/red-hat-audit-to-eradicate-problematic-language-in-its-code.

8. S. Lewis (2019). The racial bias built into photography. *The New York Times*. Available at: https://www.nytimes.com/2019/04/25/lens/sarah-lewis-racial-bias-photography.html.

9. Noble, *Algorithms of oppression*, 2018.

10. L. Peeples (2020). What the data say about police brutality and racial bias-and which reforms might work. *Nature*, 583(7814), 22–24.

11. D. Parè & P. Sengupta (In Press). Queering computing and STEM education. In: *Oxford Research Encyclopedia of Education*. Oxford.

12. T. M. Philip & P. Sengupta (2020). Theories of learning as theories of society: A contrapuntal approach to expanding disciplinary authenticity in computing. *Journal of the Learning Sciences*. Available at: https://doi.org/10.1080/10508406.2020.1828089.

13. C. E. Sleeter (2011). Becoming white: Reinterpreting a family story by putting race back into the picture. *Race Ethnicity and Education*, 14, 421–433.

14. P. Sengupta, A. Dickes, A. V. Farris, A. Karan, D. Martin, & M. Wright (2015). Programming in K–12 science classrooms. *Communications of the ACM*, 58, 33–35.

15. P. Sengupta, B. Kim, & M. C. Shanahan (2019). Playfully coding science: Views from preservice science teacher education. In *Critical, Transdisciplinary and Embodied Approaches in STEM Education*, 177–195. Springer.

16. A. Hostetler, P. Sengupta, & T. Hollett (2018). Unsilencing critical conversations in social-studies teacher education using agent-based modeling. *Cognition and Instruction*, 36(2), 139–170.

17. W. G. Sumner (1906). *Folkways: A study of the sociological importance of usages, manners, customs, mores, and morals.* Ginn and Company.

18. M. Hewstone, M. Rubin, & H. Willis (2002). Intergroup bias. *Annual Review of Psychology*, 53, 575–604.

19. G. Balabanis & A. Diamantopoulos (2004). Domestic country bias, country-of-origin effects, and consumer ethnocentrism: a multidimensional unfolding approach. *Journal of the Academy of Marketing Science*, 32(1), 80–95.

20. B. Edmonds, D. Hales, & L. Lessard-Phillips (2020). Simulation models of ethnocentrism and diversity: An introduction to the special issue. *Social Science Computer Review*, 38(4), 359–364.

21. R. Axelrod & R. A. Hammond (2003). *The evolution of ethnocentric behavior.* Presented at the Midwest Political Science Convention, Chicago.

22. Here we have used labels for each strategy based on the simulation used in: U. Wilensky, & W. Rand (2007). Making models match: Replicating an agent-based model. *Journal of Artificial Societies and Social Simulation*, 10(4), 2.

23. For the earliest use of the terms "macro-behaviors" and "micro-motives" in the context of modeling social emergence, please see: T. C. Schelling (1978). *Micromotives and macrobehavior.* Norton.

24. E. Bonilla-Silva (2019). Feeling race: Theorizing the racial economy of emotions. *American Sociological Review*, 84(1), 1–25.

25. V. M. Rios (2011). *Punished: Policing the lives of Black and Latino boys.* NYU Press.

26. S. Khalili (2017). *Caucasians on camels: Iranian American intergenerational narratives and the complications of racial & ethnic boundaries.* Doctoral dissertation, UC Irvine. Available at: https://escholarship.org/content/qt2dn161c6/qt2dn161c6.pdf

27. B. Parekh (2019). *Ethnocentric Political Theory: The Pursuit of Flawed Universals.* Springer.

28. B. Dedman (1988). The color of money. *The Atlanta Journal-Constitution.* Available at: http://powerreporting.com/color/. See also: D. Massey, & N. A. Denton (1993). *American apartheid: Segregation and the making of the underclass.* Harvard University Press.

29. J. M. Wing (2008). Computational thinking and thinking about computing. *Philosophical Transactions of the Royal Society A: Mathematical, Physical and Engineering Sciences*, 366(1881), 3717–3725.

30. Z. Obermeyer, B. Powers, C. Vogeli, & S. Mullainathan (2019). Dissecting racial bias in an algorithm used to manage the health of populations. *Science*, 366(6464), 447–453. See also: A. Alkhatib & M. Bernstein (2019, May). Street-level algorithms: A theory at the gaps between policy and decisions. In *Proceedings of the 2019 CHI Conference on Human Factors in Computing Systems*, 1–13. ACM.

31. See section 1.1.

32. P. Sengupta, A. Dickes, A. Farris, A. Karan, A., D. Martin, & M. Wright (2015). Education: Programming in K–12 science classrooms. *Communications of the ACM*, 58(11), 33–35.

33. R. Axelrod & R. A. Hammond (2003). *The evolution of ethnocentric behavior.* Presented at the Midwest Political Science Convention, Chicago.
R. A. Hammond, & R. Axelrod (2006). The evolution of ethnocentrism. *Journal of Conflict Resolution*, 50(6), 926–936.
R. L. Riolo, M. D. Cohen, & R. Axelrod (2001). Evolution of cooperation without reciprocity. *Nature*, 414, 441–443.

34. J. M. G. Van der Dennen (1995). *The origin of war: The evolution of a male-coalitional reproductive strategy*, Vols. 1 & 2. Origin Press.

35. J. G. Klein & R. Ettensoe (1999). Consumer animosity and consumer ethnocentrism: An analysis of unique antecedents. *Journal of International Consumer Marketing*, 11(4), 5–24.

36. C. D. Kam & D. R. Kinder (2012). Ethnocentrism as a short-term force in the 2008 American presidential election. *American Journal of Political Science*, 56(2), 326–340.

37. Hammond & Axelrod, The evolution of ethnocentrism, 2006.

38. Hammond & Axelrod, The evolution of ethnocentrism, 2006.

39. S. De, M. J. Gelfand, D. Nau, & P. Roos (2015). The inevitability of ethnocentrism revisited: Ethnocentrism diminishes as mobility increases. *Scientific Reports*, 5, 17963.

40. Hammond & Axelrod, The evolution of ethnocentrism, 2006, p. 928.

41. R. Boyd & P. J. Richerson (1990). Group selection among alternative evolutionarily stable strategies. *Journal of Theoretical Biology*, 145(3), 331–342. See also: Riolo et al., Evolution of cooperation, 2001.

42. Hammond & Axelrod, Evolution of ethnocentrism, 2006, p. 928.

43. See for example: D. E. Brown (2004). Human universals, human nature, and human culture. *Daedalus*, 133 (4), 47–54.

44. Parekh, *Ethnocentric political theory*, 2019.

45. See for example: S. Oishi et al. (2013). Residential mobility increases motivation to expand social network: But why? *Journal of Experimental Social Psychology*, 49, 217–223.

46. M. Pollock (2004). Race bending: "Mixed" youth practicing strategic racialization in California. *Anthropology & Education Quarterly*, 35(1), 30–52.

47. M. Emirbayer & M. Desmond (2015). *The racial order*. University of Chicago Press.

48. D. E. Hess & P. McAvoy (2014). *The political classroom: Evidence and ethics in democratic education*. Routledge.

49. As we mentioned in Chapter 1, Papert borrowed this term from Winnicott: D. W. Winnicott (1953). Transitional objects and transitional phenomena—A study of the first not-me possession. *International Journal of Psycho-Analysis*, 34, 89–97.

50. Hostetler, Sengupta, & Hollett, Unsilencing critical conversations, 2018.

51. P. H. Collins (2002). *Black feminist thought: Knowledge, consciousness, and the politics of empowerment*. Routledge.

52. Hostetler, Sengupta, & Hollett, Unsilencing critical conversations, 2018.

53. U. Wilensky & W. Rand (2007). Making models match: Replicating an agent-based model. *Journal of Artificial Societies and Social Simulation*, 10, 2.

54. R. Axelrod & R. A. Hammond (2003). *The evolution of ethnocentric behavior.* Paper presented at the Midwest Political Science Convention, Chicago, IL.

55. U. Wilensky (1999). NetLogo. Center for connected learning and computer-based modeling. Northwestern University.

56. P. Sengupta, A. Dickes, A. V. Farris, A. Karan, D. Martin, & M. Wright (2015). Programming in K–12 science classrooms. *Communications of the ACM*, 58, 33–35.

57. Source: https://demographics.virginia.edu/DotMap/index.html.

58. Hostetler, Sengupta, & Hollett, Unsilencing critical conversations, 2018.

59. See Tami's case, in Hostetler, Sengupta, & Hollett, 2018, 21.

60. Hostetler, Sengupta, & Hollett, Unsilencing critical conversations, 2018.

61. E. Ackermann (2012). Perspective-taking and object construction: Two keys to learning. In *Constructionism in practice*, 39–50. Routledge.

62. See Jamie's case, in Hostetler, Sengupta, & Hollett, 2018, 22–23.

63. M. L. Britton & P. R. Goldsmith (2013). Keeping people in their place? Young-adult mobility and persistence of residential segregation in US metropolitan areas. *Urban Studies*, 50(14), 2886–2903.

64. S. J. South, Y. Huang, A. Spring, & K. Crowder (2016). Neighborhood attainment over the adult life course. *American Sociological Review*, 81(6), 1276–1304.

65. https://www.citylab.com/equity/2012/11/where-find-diversity-america/3892/

66. See N. Montfort (2017). *The future*. MIT Press. As quoted in: A. Garcia & T. M. Phillip (2018). Smoldering in the darkness: Contextualizing learning, technology, and politics under the weight of ongoing fear and nationalism. *Learning, Media and Technology*, 43(4), 339–344.

67. J. Dewey. (1917). The need for a recovery of philosophy. In *Creative intelligence: Essays in the pragmatic attitude*, 3–69. Holt.

68. M. Emirbrayer & M. Desmond (2015). *The racial order*. University of Chicago Press.

69. G. H. Mead (1934). *Mind, self and society*, Vol. 111. University of Chicago Press.

70. Emirbrayer & Desmond, *The racial order*, 2015.

71. J. W. Loewen (2007). *Lies My Teacher Told Me: Everything Your American History Textbook Got Wrong*. Touchstone.

72. Hostetler, Sengupta, & Hollett, Unsilencing critical conversations, 2018.

73. P. Seixas (2015). A model of historical thinking. *Educational Philosophy and Theory*, 49(6), 1–13.

6 Computational Heterogeneity and Teacher Voice

6.1 Introduction

A persistent theme in this book is how heterogeneity in representational and epistemic work is key to understanding how coding can become an integral part of *science as practice* in the classroom. In this chapter, we focus on the relationship between heterogeneity of computational utterances and teacher voice in an elementary science classroom. Teacher education is beginning to receive attention in educational computing research, in the forms of both teacher professional learning[1] and preservice teacher education courses.[2] However, longer term investigations of teachers *developing their voice* in the context of supporting their students' computational utterances in science classrooms have barely received attention. From a phenomenological perspective, we believe that this is an essential and yet largely ignored aspect of educational computing.

We followed an elementary teacher, Emma, and her students over the course of two years in her science classroom. We present a phenomenological account of how the experience of computing in this classroom becomes intricately woven with the practice of doing science as Emma *manages* the heterogeneity inherent in computational and scientific modeling, through establishing and refining classroom norms. In terms of radical reflections that are characteristic of phenomenological inquiry,[3] this chapter reveals how designing heterogeneous forms of computational utterances and iteratively developing disciplinarily grounded classroom norms (e.g., sociomathematical norms) to bring about disciplinary coherence across these representations play important roles in Emma's (and her students') finding their computational voices. In many cases, these utterances extend far beyond the computer and the programming language and yet bear deep conceptual connections to computational representations within the programming language. Over the course of two years, we showed how Emma progressively recognized the value of representational heterogeneity in computational design and, at the same time, anchored her classroom instruction through framing computational work as the normative design of mathematical measures.

6.2 Background

6.2.1 Computational Heterogeneity and Science Education

An important critique of technocentrism can be found in diSessa's book *Changing Minds*[4] in which he has argued that computational literacy can be characterized as an interplay between material, social, and cognitive dimensions in which "valued intellectual ends" are achieved through "widespread, patterned deployment of skills and capabilities in a context of material support" (page 19). DiSessa thus reminds us that it is important to distinguish between *intelligence* and *literacy*. Focusing solely on computational thinking as *cognitive intelligence* and/or *material intelligence* would restrict the scope of experience to reasoning about computational abstractions and/or the intelligent use of technologies or tools. But for any of these forms of "intelligences" to become literacies, they must come to be seen as *infrastructural* to a society's communicative practices.[5] If one views the science classroom as a community in which computational literacy and scientific literacy develop hand in hand then we must take into account the interactions between cognitive, social, and material dimensions that construe such literacies.[6]

This image of computational literacy is well aligned with the notion of practice that has greatly shaped science education.[7] In one sense, practice brings into view the essential role that communities play in shaping individual experiences.[8] Latour's analysis of how sharing scientific representations among scientists shapes the development of scientific knowledge is also an example of this view. In another sense, a practice-based perspective of "doing science" involves viewing scientific work as a dance of agency between ideas and the physical world.[9] This implies a deep, inextricable intertwining of epistemic work with representational work. The broader community of scientists is present here, too—albeit implicitly—in distinguishing between and synthesizing across different genres of representations that have been valued historically by the community of scientists as appropriate representations in similar contexts.[10]

Adopting this *practice*-based perspective of science education leads us to think carefully about two things. First, it draws our attention to modeling as a key form of activity that students and teachers must engage in—modeling has indeed been regarded as the *language* of science.[11] We introduced this argument in chapter 2 by highlighting the relationships between modeling in science and computational thinking and by showing how viewing coding as a language can help us imagine the continuities between them (section 2.4). We argued that heterogeneity is at the heart of the practices of both scientific modeling and design, and therefore should also be positioned centrally in our imaginations of computing. This now brings us to the second point: a key finding in the science education literature is that heterogeneity is inherent and essential in learning to model science. Studies of how children develop scientific expertise through iterative cycles of modeling throughout the entire academic year, and often over multiple years, demonstrates how learning modeling involves creating, comparing, and critiquing different forms of inscriptions, and using multiple forms of material means to create physical representations and models.[12] As we discuss in this chapter, this is also true in the context of using computational programming and simulations as the medium of modeling in the classroom. The work in which the students and teachers engage within such contexts is far richer than merely programming the computer using the programming commands specified in the language—what we more broadly term in this book *device-level engagement*.[13]

Focusing on heterogeneity, in turn, is important for two reasons. First, when different inscriptions and representational systems that highlight different aspects of the same phenomena are brought into contact with each other—a phenomenon that Latour[14] termed *circulating reference*—their juxtaposition *amplifies* the conceptual field at play.[15] In the classroom this creates opportunities for negotiating different interpretations of the same phenomenon and supports different ways of noticing and representing the phenomenon. The second reason, as Lehrer argues, is that by engaging in modeling, children develop familiarity with the uses and trade-offs among different systems of inscriptions. This means that learning about X becomes deeply intertwined with learning about representing X in light of particular goals, a view that is deeply synergistic with the practice-based perspective of science. One could further argue that through such experiences, what develops is a form of expertise that diSessa[16] termed *metarepresentational competence*—a practice that encompasses the ability to select and design optimal external representations, design and use novel external representations, and invent new representations as necessary. While it is a competence that children bring to any educational setting, developing scientific expertise also involves bootstrapping these forms of competence and refining these abilities in disciplinarily authentic ways. For example, scientists grapple with modeling in terms of qualities such as conciseness, completeness, and precision. DiSessa argued that these qualities are also natural parts of students' considerations in designing their own versions of such representations, and thus can be leveraged through careful pedagogical design.[17]

However, our goal here is to move beyond accounts of individual competence focused solely on bootstrapping the abilities of the learner, as well as techno-centric accounts of the "charismatic"[18] power of the computational media. Instead we highlight how heterogeneity of representational forms and discourse is central to integrating computational media in the scientific work that happens in classrooms in which teachers have limited or no prior experience with programming. We highlight how heterogeneity is central from the perspectives of listening to, supporting, and developing teacher voice. Unfortunately, this is an area in which there is a great scarcity of research in educational computing, and this chapter is an effort to at least shed some light on this issue.

6.2.2   Discourse and Teacher Voice in Educational Computing

As we have argued in chapter 1 (see section 1.4), the constructionist origin of educational computing is deeply rooted in agnosticism about stereotypical images of didactic instruction in schools. Papert[19] argued against instruction-ism, noting that deliberate teaching is antithetical to Piaget's observations of children's learning without explicit instruction by others. He argued that children must be positioned to be the architects of their "own intellectual structures with materials from the surrounding culture."[20] The goal of programming, in Papert's vision, was to position children as epistemologists through engaging them in debugging their own work. This view also limits child computer inter-action to device-level engagement. In addition, as we have argued in section 1.4, ignoring the role of adult assistance and adopting child-centered and tech-nocentric views can have serious implications, as we may not be able to under-stand the crucial role that infrastructural supports, including teaching, play in classroom integration of computational modeling. This becomes particularly important in the context of working with marginalized students who typically are not provided with adequate pedagogical support and opportunities in com-puting[21] and instead have been *systematically*[22] labeled as "failures," a label that is hard to overcome.[23]

Coding in science classrooms can then be seen as quite complex and even challenging, because it involves synthesizing disciplinary practices of comput-ing and science. At the same time, there are deep synergies between practices that are central to coding—for example, designing computational representa-tions, defining patterns, generalizing from instances, parameterization, and so on—that are also central to the practice of scientific modeling.[24] We view teaching as *essential* in bringing about this synthesis. In Bakhtinian terms, teaching is not merely a form of redoubling—that is, reproducing two disciplines (computing and science) in the classroom—but rather it is a form of translation across representational genres, through which disciplinary practices are reframed axiologically.[25] That is, what is *valued* or *valuable* in coding is recast by teachers in light of their curricular and instructional needs in the science classroom. Coding is thus reframed as designing mathematical measures in the science classroom,[26] as we also find in Emma's case.

A few studies have investigated the microdynamics of teaching coding in K–12 science and STEM contexts. These studies tell us that teaching in such classrooms involves dealing with the programming syntax as well as managing conceptual difficulties in understanding scientific and technical terminologies, mathematical relationships, and disciplinary practices for representing these relationships.[27] Perspectival heterogeneity is fundamental to computational modeling, and as chapter 3 shows, even apparently simple tasks—such as interpreting what a computational modeling activity is asking students to do—require complex negotiations between the myriad perspectives at play (the code, the real world, relevant scientific terminologies, mathematical measures, and so on). In such contexts, focusing conversations around how programming commands are being used can be helpful in some cases,[28] whereas doing so may also turn the teacher's attention away from other important elements of the students' experience that would benefit from being elaborated and acted upon.[29]

So, given this uncertain and understudied terrain, how do teachers shape the culture of computing in science classrooms? To answer this question, we present an in-depth case study of how Emma, an elementary school teacher, designed and organized instruction in order to manage the heterogeneity that emerged in a classroom in which computer programming and modeling were used as a medium for *doing* science on a regular basis (two days a week), throughout the entire year, for two consecutive years. We illustrate how emphasizing mathematizing and measurement as key forms of learning activities helped Emma meaningfully frame coding as a language of science. Specifically, Emma accomplished this by designing and incorporating multiple forms of modeling—similarly to Latour's notion of *circulating references*—and by supporting the iterative development of classroom norms for assessing the "goodness" of computational models. But before we dive into Emma's world, in the following section we present a review of the relevant literature in mathematics and science education to help us gain an understanding of how the kind of norms that Emma developed can shape the (micro)culture of modeling in elementary classrooms.

6.2.3   Norms and the Culture of Modeling

Education researchers who have studied how cultures of modeling develop over long periods of time (ranging from several months to a year) in mathematics and science classrooms have demonstrated that the iterative development and refinement of collective, that is, classroom-level, *normative* modeling practices can greatly deepen students' conceptual and representational work.[30] Science educators have shown that the question of what counts as a "good" model also needs to be normatively established in classroom instruction in order to deepen students' engagement with scientific modeling in elementary grades, and that these norms also follow similar shifts toward deeper disciplinary warrants over time.[31] An important finding that emerges from the studies conducted by Lehrer and colleagues is that in such science classrooms, mathematics becomes a meaning-making lens through which the natural world can be systematized and described through cycles of iterative development and refinement of normative understandings of what counts as a good model. In such classrooms, epistemic work develops and deepens through the social construction of disciplinary knowledge.[32] The authors argue that an emphasis on measurement, including aspects of measurement such as precision and error, and normatively guided model refinement help students move beyond a focus on superficial features of the target phenomena to modeling "unseen" relationships between variables and underlying mechanisms.[33]

As we have reported elsewhere,[34] the specific genre of norms we found to be relevant to our work has been termed *sociomathematical norms* by Cobb and his colleagues.[35] Sociomathematical norms refer to the normative aspects of classroom actions and interactions that are specifically mathematical. They emerge through interactions with and conversations about mathematical objects and are ascribed social value by the members of the classroom. These norms shape and, at times, regulate classroom discourse. Early work by Cobb and colleagues showed that teachers initiate and guide the sociomathematical norms characterized by explanation, justification, and argumentation.[36] Similar to their work, our focus is on the perspective of the teacher (Emma), who initiated these norms on her own accord, without any prompting by the researchers.

Cobb and colleagues noted that "what counts as an *acceptable* mathematical solution" is an example of a fundamental sociomathematical norm. Furthermore, this norm typically originates as a *socially* defined norm, and shifts over time to a *sociomathematically* defined norm. They identified key characteristics of this shift: first, the shift from social to sociomathematical is grounded in deepening of mathematical warrants. That is, as this shift happens, one should expect to see deeper mathematical discourse in the classrooms because these warrants are indicative of the emergence of shared understandings of what counts as disciplinarily acceptable mathematical practices. Second, discussions around student explanations play important roles in the shaping of sociomathematical norms. Yackel and Cobb present cases in which what is "taken as shared"[37] is established during class discussions in which teachers and students negotiate what counts as acceptable mathematical solutions. In the context of adopting computational modeling and programming in the science classroom, particularly from the perspective of teachers who themselves do not have prior experience with programming, we see that sociomathematical norms can play a powerful role. In this chapter, we see how as Emma develops and refines sociomathematical norms along with her students, she connects coding with science by reframing programming as a means to design mathematical measures and brings together heterogeneous but complementary forms of modeling (e.g., physical, embodied, and computational modeling).

However, given that these norms are often teacher-initiated, it is also important to look at how these ideas and opportunities are taken up by students in their work. Positioning plays an important role in this aspect—for example, when students are positioned in collaborative work such that they can *offer* explanations, they can reexamine their initial understandings by restating their ideas in ways that would be comprehensible to others, and by having their ideas made available to others for examination and critique.[38] Yackel and Cobb[39] also found that in creating a culture of sharing mathematical explanations, students' explanations became objects of reflection for other members in the class. Another way in which teachers may shape students' positioning is by supporting interpersonal dispositions that position students as contributors as well as listeners. For example, Lampert[40] reported a study in which she created a classroom culture to help students be both "courageous" and "modest" in their mathematical activity. By being courageous, students took risks in sharing information that they were not really confident about, and by being modest, they were open to suggestions and critiques from others. Lampert's work showed how such dispositions can productively support students to deeply engage with the discipline.

What stands out from our partnership with teachers attempting to integrate coding with their year-long science curricula are their efforts to establish normative criteria and questions in order to advance the computational work as well as the positioning of students within the activities. Here, we present an illustrative example following Emma over two successive years, and situate Emma's voicing as *(re)framing* coding as *mathematization* in the science classroom, through which certain forms of computational work are *taken as shared* as acceptable forms of scientific representations. Furthermore, how Emma positions her students as contributors and listeners has interesting differences across the two years that we followed her and her students, and we look at how these shifts differently shape the heterogeneity in the works of her students. We also discuss how this in turn indicates shifts in epistemological stances on Emma's part, and has implications for the development of teacher voice in the context of coding in STEM classrooms.

6.3   The Setting: Emma's Classrooms in Year 1 and Year 2

During Year 1 of the study, Emma was teaching 3rd grade students in a 99% African-American public charter school located in a large metropolitan school district in the Southeastern United States. Her class had 15 students—14 African-American and one Latino—and all of them participated in this study. During Year 2, Emma was assigned to a 4th grade classroom, and seven of her 15 students from 3rd grade moved with her to 4th grade. In each year, our focus was to observe and understand how Emma used coding and modeling with ViMAP twice a week throughout the academic year (roughly eight months, each year) as part of her regular, district-, and school-mandated science lessons and help her, as needed, in accomplishing what *she wanted to do* with ViMAP and coding in her science lessons. We met weekly with Emma and discussed her objectives and her students' progress. During these meetings we worked with Emma on co-designing classroom activities, with Emma clearly leading both the co-design process and the classroom instruction.

6.4   The Learning Activities

During both Year 1 and Year 2, the learning activities were divided into three phases: Phase I (Geometry), Phase II (Kinematics), and Phase III (Ecology). In this chapter we share vignettes from Phases I and II (Geometry and Kinematics) in Year 1 and Phase II (Kinematics) in Year 2, and trace the development of normative, mathematically grounded criteria for what counts as "good" ViMAP models of motion. Following our earlier work, instruction during each phase blended complementary forms of modeling, including embodied, physical, and computational modeling, to generate scientific artifacts and discretized motion into steps to scaffold understanding of motion as a process of continuous change (e.g., aggregating steps to produce constant motion). The learning activities were also designed to support the invention and interpretation of mathematical measures of distance and provide opportunities for the classroom teacher to reframe programming as a mathematization of motion. Figures 6.1 and 6.2 summarize the relevant learning activities across both years.

In Emma's work, there was a particular tension between supporting students' exploration of the representational palette offered in ViMAP and her emphasis on the production of canonical representations valued or mandated institutionally and from a curricular perspective. For example, Emma explained to us at the beginning of Year 1 that for her, graphs were the most important "output" of ViMAP models, because they would help her connect students' ViMAP models with representational forms that are mandated by the curriculum and frequently assessed on standardized tests. Throughout the study, Emma maintained that her goal was to help students develop computational models that produced graphs of change over time. Of course, students could decide to use ViMAP to create more imaginative representations, and we (the researchers) would often show Emma examples of how students might use other ways of creating models that were more expressive. Tables 6.1 and 6.2 list the programming commands (that is, command blocks) within the ViMAP commands library that were available to Emma and the students during the study. Emma, however, remained committed to emphasizing graphing as *the connection* between coding with ViMAP and her regular instruction in mathematics and science. Due to this commitment to curricular mandates, Emma's instruction with ViMAP in Year 1 emphasized multiplicative reasoning and sides and perimeter in geometry and learning to understand and represent motion in terms of forces. As our partnership progressed in Year 2, we noticed that Emma began to value more explicitly the epistemic and representational diversity in her students' works. The vignettes we highlight in our analysis demonstrate how Emma supported greater heterogeneity in students' representational and epistemic work by emphasizing rather than de-emphasizing students' interpretive decision making, and by creating opportunities for the class to reason about trade-offs about what the models could and should show.

Phase	Activity	Description
Phase I	Shape Drawing	Students work in pairs writing rules and creating ViMAP programs for drawing squares, triangles, and circles.
	Regular Polygons	Students derive a formula for finding the exterior angle of regular polygons (# of sides/360) and use that formula to draw any regular shape in ViMAP.
	Congruent Shapes	Students program congruent shapes in ViMAP
	Perimeter	Students use ViMAP s graphing function to find the perimeter of geometric shapes. Students discuss how ViMAP graphs are "unfolded" polygons.
Phase II	Leaving Footprints	Students leave ink footprints on banner paper.
	Generating Measures	Students iteratively develop, apply, test and refine a measurement of distance termed a step size.
	Collecting Step-Size Data	Students use the step size measurement convention to measure their personal step sizes.
	Modeling Step Sizes in ViMAP	Students model their personal step sizes in ViMAP. Total-distance graphs and predictions using ViMAP's grapher are generated and discussed.
	Modeling Motion as a Process of Continuous Change	Students model motion scenarios in ViMAP and check the validity of those models using ViMAP's grapher and the total distance equation.

Figure 6.1
Summary of Phases I and II learning activities (3rd grade, Year 1)

Phase	Activity	Description
Phase II	"Constant Speed" Robots	Goal is for students to develop understanding of speed as a rate of the distance traveled in a unit of time, including cycles of model sharing and revision. Small groups of students first conduct physical investigations and measurement activities using a Lego NXT robot's motion and model its motion using both ViMAP and embodied, physical, and paper-based representations.
	Constant Acceleration and Gravity	Contexts are acceleration down a ramp and free fall. Small groups of students transform references across physical modeling, video analysis, ViMAP to find ways to measure and model continuous changes in speed.
	Friction	Students model processes of "slowing down" for Matchbox cars on different surfaces.

Figure 6.2
Summary of Phase II learning activities (4th grade, Year 2)

6.5   Data and Analysis

Data for this work comes from informal interviews with the participants, video recordings of class activities and discussion, student artifacts (e.g., student representations, activity sheets, and ViMAP models), and daily field notes. During class, the researchers and the classroom teacher conducted informal interviews during opportune moments while the students were engaged in single, pair, or small-group work around modeling and representational activities. Classes were video-recorded, and student-created artifacts (ViMAP models, written work, and so on) were also collected.

We present the analysis in the form of explanatory case studies,[41] using the constant comparative method.[42] Case studies are well suited as a methodology to answer *how* and *why* questions. We find this to be a good fit because our goals here are twofold. First, we seek to identify *how* and *why* Emma designed and used circulating references in order to appropriate coding as part of her regular classroom activities. Second, our goal is to illustrate the *process* through which Emma and her students developed sociomathematical norms, which includes answering *how* the development of these norms shaped the students' interactions with ViMAP and other modeling experiences, and *why* these norms were deemed useful by the teacher.

It is important to note that both of the "lenses" of looking at teacher voice—circulating references and sociomathematical norms—emerged from our analysis through the constant comparative method. Furthermore, these lenses *in use*—that is, as experienced by Emma and her students—emerged to be deeply intertwined with one another. This means that the use of circulating references was deeply tied to how sociomathematical norms were being developed and becoming taken as shared in the classroom. Furthermore, the comparison across Years 1 and 2 reveals how Emma's voicing of sociomathematical norms shifts toward accommodating and valuing a greater role of her students' interpretive moves. In doing so, Emma's vision of *what counts as a good computational model* also broadens, while still being deeply grounded in her and her students' use of circulating references. We also asked the following question: what work does this form of teaching do for growth in children's use of computational abstractions? In order to shed some light on this, we present a classroom-level analysis of students' ViMAP code and models in terms of the quality of their code during each year.

Table 6.1
ViMAP command blocks for movement and measurement

Movement Blocks	*Measurement Blocks*
set step size [variable]	
set heading [variable]	
right [variable]	
left [variable]	place measure point
go forward	clear measure points
go backward	start over measuring
set xy [variable] [variable]	label [variable]
set xy random [variable] [variable]	
heading [variable],	
set [variable] plus/minus [number]	
set [variable] equal to [variable]	

Table 6.2
ViMAP command blocks for drawing and control

Drawing Blocks	*Control Blocks*
set [variable] equal to [number]	set [variable] = [current value] + [number]
pen down	set [variable] = [variable]
pen up	If [condition] Then [commands] Otherwise [commar
stamp	Repeat [number of times]
go invisible	If [variable] [<, >, =] [variable]
go visible	If [variable] [<, >] [number]
change [current shape] to [new shape]	If [variable] [<][variable]

6.6   Year 1: Voicing Code Using Circulating References

6.6.1   Vignette 1: Squares, Loops, and Multiplicative Reasoning

We met with Emma for the first time in September of the first school year. During this meeting, Emma explained that she was interested in exploring the connections between ViMAP and her math curriculum. Along with the authors, Emma then explored drawing ViMAP shapes (e.g., squares) through creating simple programs, and how she could create graphs that could "measure" the size of each side of the square as well as the perimeter of the shape as it is being drawn by the Turtle. She wanted to introduce this activity to her students as a way to learn programming ViMAP while also learning mathematics. So, Phase I activities began by introducing students to ViMAP through shape-drawing activities. The goal was for students to work in pairs to create rules for drawing squares, triangles, and circles (e.g., go forward 5 steps, turn right 90 degrees, go forward 5 steps, and so forth) and translate those rules into ViMAP programs. In each pair, Emma wanted one student (the "drawer") to try to draw the square through embodied movement (stepping on the floor), and the other student (the "rule maker") to try to record in simple English what the "drawer" was doing. Students would then switch roles and compare notes, finalize their rules for drawing a square, and finally, try to create these rules in ViMAP using commands from the ViMAP command library.

During the first two days of instruction, Emma requested that one of us lead the instructional activities that involved ViMAP, while she walked around the room interacting with pairs of students and leading embodied modeling activities in which she would ask students to "draw shapes with [their] bodies" through movement. Emma's hesitation with ViMAP was evident as she referred most questions about the ViMAP commands to us. However, one key interaction between Emma and one of her students, Kendra, changed how Emma positioned herself in relation to supporting students in making with ViMAP. During the first shape-drawing tasks Emma observed how computational control structures such as a repeat loop could be used as an instructional tool for teaching multiplicative reasoning.

Working with Kendra to draw a square, Emma asked Kendra to compare the number of commands, or rules, needed to make a square using the repeat loop to the number of commands needed to make a square without a repeat loop. Kendra noticed that eight rules were needed to make a square without a repeat loop, whereas only two were needed with a repeat loop (Figure 6.3). Emma pushed Kendra to think about the number of repeat loops (four) in her program. Kendra realized the two ViMAP programs for generating a square were equivalent, noting that "four times two equals the eight commands." Emma

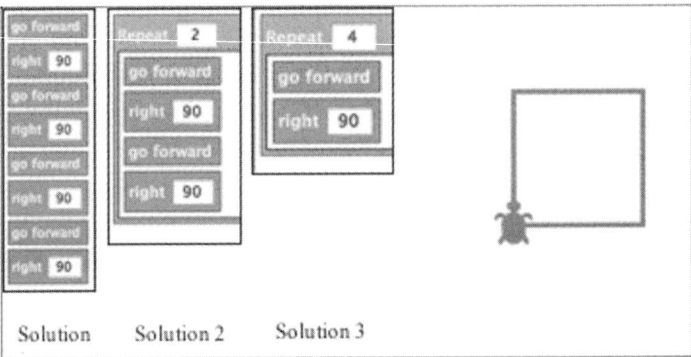

Figure 6.3
Kendra's three ways to make a square

challenged Kendra to think of another way to "make a square" and Kendra produced a ViMAP program using a repeat two with four commands in the repeat loop. Kendra's three ways to make a square are shown in Figure 6.3. Immediately after class, Emma approached Amanda and Amy and expressed her realization that there was "so much math" in ViMAP and communicated her interest in finding more opportunities to utilize ViMAP to teach mathematics. Among the ideas discussed were enhancements to a future lesson on regular polygons and investigating the differences between circles and other closed, regular shapes.

This vignette represents an important moment for Emma in terms of her beginning to *voice* programming. She was visibly excited at the end of the class because—and it is not a far stretch to connect the dots here—she had now *experienced* how programming could become reflexive with her curricular needs in mathematics. Learning to use loops made multiplicative reasoning explicit to her students, and learning about shapes and angles in geometry was intricately woven in with thinking about Turtle graphics. In a happenstance meeting with Pratim in a local cafe that weekend (two days after this class), Emma reiterated her excitement, further noting that several of her students who would otherwise not speak during math instruction were now sharing their mathematical thinking with ViMAP. Kendra was one of these students who was mostly silent in her mathematics class, and Emma noted that she was "seeing a different Kendra with the square." Emma also explained that she could now connect her previous experience teaching first grade mathematics with her experience of "doing math with ViMAP," and had already begun planning her next activity. She commented:

I can bring back the body—I would ask kids to make the shapes with their bodies, and I would really like to do a lot of (multiplication) tables with them.

Pratim told Emma that historically, Logo researchers had found that the forms of activities that she proposed work quite well with young students. Emma requested that she would like to lead the lesson planning for the next class (three days after the meeting in the cafe). This was the first time Emma voiced her intention to lead the design of learning activities, and would continue to do so for the rest of the academic year. Throughout the year, the emphasis on "doing math" with ViMAP persisted in the activities: the connections with mathematical representations and reasoning had positioned ViMAP as *valuable* and *useful* to Emma, who continued to use ViMAP by designing mathematical investigations for her students in her science lessons.

6.6.2   Vignette 2: Loops, Turns, and Closed Shapes

In the next class, students began their investigations of closed, regular shapes by working in pairs to write rules for drawing a circle. Drawing a circle *body-syntonically*[43] involves moving a little, turning a little, and then repeating these two actions until we reach where we began our journey. The programming commands resemble these embodied actions, that is, thinking about how much the Turtle should turn every step, represented by the variables "right-turn" and "left-turn," the length of forward movement, represented by the variable "step-size," and the number of steps, which is represented as loops. What is noteworthy here is that Emma, who was certainly new to programming and was not aware of Papert's writing or the broader Logo research, intuitively arrived at *body syntonicity* because she recognized that embodied activities can make geometry intuitive for her students while at the same time providing students with a "concrete" (in her words) experience for understanding ViMAP commands.

Emma introduced circles and other regular polygons as "closed shapes" because the Turtle must come back to the starting position in order to complete drawing the shape. Students had already drawn a square, so Emma seeded the circle-drawing activity by instructing the drawers in each group to take small steps forward (short step size), while turning 90 degrees four times, a familiar turn angle from square building. Student pairs reasoned that, as they had rotated a full 360 degrees, they had successfully "made a circle" with their bodies.

Working with one student pair, Jayla and Keone, Emma asks them to think about their turn angles: "If we turn 90 degrees is that too far or far enough?" Jayla and Keone are both confident that turns of 90 degrees are sufficient, using the reasoning stated earlier. Emma then introduces an "important" characteristic of circles as a "closed shape with no straight edges" and with her attention

Figure 6.4
Jayla and Keone walk a square and a circle.

still focused on Jayla and Keone, bends down to draw a circle and square on the floor of the classroom using a dry-erase marker. Emma asks Keone to walk the square first, then as he prepares to walk around the circle Emma tells him "Now, I want you to walk it, and notice *what you do* when you walk it." Keone carefully traces the outside of the circle, setting one foot in front of the other as he walks (Figure 6.4). "Oh, I curved!" he exclaims as he works his way around the circle. Emma draws a second square around the circle to help Jayla and Keone see the difference between shapes in terms of "smoothness" of the circle and the differences in turn angle, before requesting that both of them try to express what they have just discovered in more mathematical terms.

Emma next pulls the entire class together and asks her students to "tell her something that is different" between a circle and a square. Nearly unanimously, students commented that squares had "corners" while circles were "smooth." Emma emphasizes that circles have *no straight lines*, an important idea to which she would return later in the session. She then charges her students with programming smooth circles using ViMAP. Emma wanders around the class looking at students' computers and asks a student, Kenya, to share her program for drawing a circle with the class. Kenya displays her program and Emma reads aloud each line of her code, asking her to explain what each command does. For example, she asks, "What does 'go forward' do? Why do we have that in there?" A student, Darien, answers Emma's question: "Go forward makes *you* move forward." Emma wonders what the Turtle would do if there was "no go forward," and asks Kenya to demonstrate the Turtle's actions using her body (she asks Kenya, "show me with your body"). As shown in Figure 6.5, Kenya responds to Emma's questions by spinning around in a circle.

Kenya's code (Figure 6.6), while giving the appearance of a circle with no straight edges, revealed itself to be more problematic when Emma requested

Kenya delete the stamp command and rerun the simulation so that everyone could see the circle better. After removing the STAMP command, the circle was no longer a "perfect" circle, being neither a "closed shape" nor lacking "straight edges." Emma remarks to the class, "Maybe the stamp command hides some things," and asks how they could help Kenya "fix" her circle. A student, Darien, recommends decreasing the turn angle, which "opens" the circle even more. Emma re-asks the question, focusing students' attention to the turn angles. "What angle do we need to put in to get rid of the straight lines?" Some students suggest getting rid of the color command and Emma responds, "This is important: circles do not have straight lines. Is (the) *changing color* command going to help us fix our lines?" A bit of frustration evident in her voice, Emma says to the class, "I don't know if I can show this to you all, it's not connecting." It is becoming clear to Emma that the shape not "closing" is indicative of her students not understanding the relationship between angles, step size, and repeat. She turns to Amanda and Amy and states, "I just want them to really . . . it's such an abstract thought." Emma thinks for a few seconds and then launches another embodied modeling activity—a form of activity that she had previously referred to as "concrete"—to illustrate the relationship between angles, step size, and repeat (number of steps): "circle races."

Figure 6.5
Kenya walks a circle: Kenya represents what would happen if the Turtle did not go forward by standing in one place and turning around in a circle.

Emma asks Kenya to join her in "circle racing." She tells Kenya that they both will "repeat 10 times," that is, move for the same number of steps but each turn a different amount at the end of every step. She asks Robin, another student, to help her keep track of their steps. Emma selects an angle of 25 degrees, while Kenya turns 15 degrees. Counting off 10 "big steps," Emma and Kenya walk their circles (Figure 6.7). Emma's circle fully closes, while Kenya's does not. Emma asks the class what is different between the two circles, reminding the class that both she and Kenya took the same number of steps and (approximately) the same size steps. One student states that the angle was different, to which Emma responds "Our angles were different, the way we turned." (Emma turns back and points to Kenya's ViMAP program.) "So when we made this angle smaller, we needed to do what?" Another student, Martin, suggests "adding steps." Emma asks the class how they add steps in ViMAP ("we're adding more steps by changing the . . . "), and a student, Darien, recommends Kenya increase her number of repeats from 17 to 21. Running the program with these values for "repeat" reveals a more "closed" circle that has fewer straight edges.

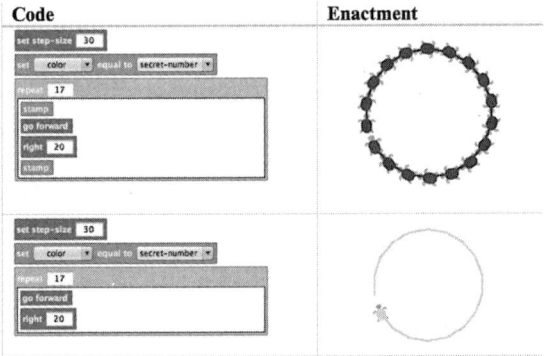

Figure 6.6
Kenya's code, with and without "stamps" of the Turtle's position. Top image shows code snippet and simulated output using the STAMP command. Bottom image shows code snippet and simulated output without using the STAMP command.

In the next few classes, Emma used ViMAP to amplify two mathematical ideas on which she was focused on the basis of her district-mandated curriculum: polygons and perimeter (Figure 6.8). Phase I thus culminated in two key activities: mathematically describing regular polygons using ViMAP by exploring the relationship between turn angles, number of sides of the polygon, and the "repeat" command; and using the ViMAP graphing functionality to connect the shapes drawn by the Turtle to mathematical measures such as distance and perimeter. In each activity, Emma was able to highlight the importance of multiplicative reasoning. Similarly to what we have seen so far, Emma continued to interweave her references to the ViMAP commands, relevant mathematical ideas and properties, and embodied modeling. The pattern of the activities was also similar to what we have seen so far: Emma would begin her class with a *less than perfect* scenario as an object of reflection for the entire class—for example, an imperfect ViMAP program for generating a nonagon. Students would then engage in a class discussion led by Emma, and then students would "fix" the ViMAP program. Emma would visit students around the class during this part of activity, and when she noticed difficulties encountered by multiple students, she would ask them to debug their work, as needed, through public embodied modeling activities similar to those described previously.

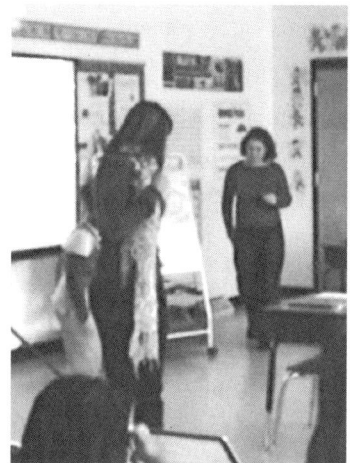

Figure 6.7
Left, Kenya and, *middle*, Emma "circle racing" in front of the class; *right*, Amy is observing.

Figure 6.8
Investigations of perimeter and polygons in Emma's classroom. The top-left shows a paper-based representation of the ViMAP shape and graphs, and the top-right shows a multiplication table that demonstrates the multiplicative relationship between the number of sides of a regular polygon and the exterior turn angle of the ViMAP Turtle.

Emma, Amanda, and Amy created one additional inscription on a classroom easel: a chart that showed how the product of the number of sides and the turn angle for each closed shape (regular polygon) was 360 degrees (Figure 6.8). Emma referred to this chart throughout students' final shape-drawing tasks, particularly when she noticed difficulties programming in ViMAP that could be resolved through multiplicative reasoning.

6.6.3   Vignette 3: Mathematically Grounding Scientific Code

Following work in geometry, Emma expressed a desire to focus the next set of activities (Phase II) on motion and graphing. She wanted to maintain an emphasis on "concrete" activities—that is, activities that involved embodied modeling by students—but also wanted to more fully incorporate the graphing functionalities within ViMAP. In conversations during weekly meetings, she explained that reasoning about units of measurement and interpreting trends in graphs of change over time were critical for her students' success in the state-mandated assessments and she wanted to use ViMAP to engage her students in classroom investigations that would directly forward this agenda. In the following vignette, we discuss *two episodes* in which Emma frames coding as an activity in which students iteratively investigate the following question: *what counts as a good model of motion?* We continue to maintain our focus on circulating references, but we also see how Emma's attempts to frame coding as mathematization leads to episodes that bear some resemblance to Cobb and his colleagues' work on sociomathematical norms.

6.6.3.1   Episode 1: Emphasizing "concreteness" and "accuracy"   In January of the new year, the transition from mathematics to motion involved students modeling patterns of movement using animal footprints as well as their own movement. Using the richly illustrated children's book "Wild Tracks! A Guide to Nature's Footprints"[44] to anchor class discussions, when Emma asked students to make inferences about animal movement by looking at the pictures and photos of their tracks, she noted that students' noticings of the animals' movements lacked "concreteness." For example, students mentioned that animals were walking or running, without being specific about any other characteristics of motion that would be valued from a curricular standpoint. Being "concrete," for Emma, meant being able to "measure mathematically." In conversations with the authors, Emma noted that her goal was to "concretize" the notion of step size as a measurable object that can be used to make scientific inferences about motion, instead of seeing step size in ViMAP only as a mathematical tool for drawing geometric shapes. Pratim, Emma, Amanda, and

Amy considered several ways for students to measure their own motion and ultimately decided to have students "stamp" their own footprints. The authors brought large banner paper and washable ink. Emma directed her students to stamp their footprints by dipping the soles of their shoes lightly in the ink, and then walking as they usually would on the chart paper.

At the end of the "stamping" activity, Emma asks her class the following question: "What is a step size? And if we were to measure one, where would we begin measuring and where would we end?" For Emma, as she noted later, it was important to ask both questions. The class offers three options for measuring step sizes: heel-to-toe, heel-to-heel, and toe-to-toe. Emma draws these three options on the whiteboard (Figure 6.9) and requests her students select the "best" option. At this stage, selection of the "best" step size measurement convention was primarily a social endeavor. Over half the class (majority) selected the heel-to-toe measurement convention because it was "the biggest" or because their "friend voted for it." This is evidence of what Cobb and colleagues have termed a socially defined mathematical norm—that is, the normatively accepted rationale for deciding which mathematical measure is the most appropriate results from a social decision without mathematical warrants.[45]

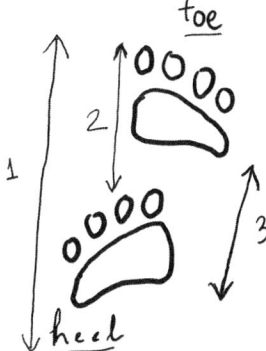

Figure 6.9
Student ideas for measuring step size: Option 1 measures heel-toe, Option 2 measures toe-toe, and Option 3 measures heel-heel. Image re-created from Emma's whiteboard drawing.

During initial negotiation of the "best" step size measurement convention, Emma recognized that the convention the students had selected was not, to use her words, an "accurate" measure of total distance traveled because it produced an overlap, effectively measuring portions of the distance twice. For

Emma, an "accurate" measure meant being able to use the measure to describe exactly how much distance was traveled, whereas students' initial measuring conventions had built-in redundancies. Even as Emma mentioned the redundancy built in the heel-to-toe measurement, her students remained happy with their decision of the "heel-to-toe" measurement.

For Emma, the redundancies meant that the chosen convention would be unusable as a measure. She noted to Amanda: "They (the students) need to understand that this [their measurement convention] won't work." But this was also an uncertain space for Emma—she further noted that she was unsure whether to "bring in ViMAP." Her emphasis on concreteness guided her next pedagogical move, as she decided to ask her students to further problematize their measures using their step sizes as "concrete" referents. Asking students to return to their stamped step size charts, she shifted the objective of the measurement to an aggregation of steps, rather than focusing only on a single step. Emma noted to the authors that her goal was to break down continuous motion into a series of discrete steps, which could then be reaggregated first through embodied action (walking) and, eventually, computational modeling (ViMAP).

What happens next in the classroom has been described in some detail elsewhere in terms of highlighting the establishment of sociomathematical norms.[46] Here we present a brief recount because it will help us "see" how the establishment of sociomathematical norms was also aligned with circulating references. In order to productively disrupt the students' acceptance of the heel-to-toe convention, Emma asks her students to return to their stamped footprints and measure their total distance traveled in two different ways: first, using a ruler, a yardstick, or a measuring tape between the first and last steps, and then using the heel-to-toe measurement convention. The latter would involve measuring each step and then adding them to calculate the total distance. Upon completing the activity, students publicly reported their findings and noticed that the two distances "didn't match," even though they had predicted that they would.

Emma then suggests to the class that "maybe [they] need to find a measure that [i]s *mathematically accurate.*" It is important to remind ourselves here that Emma's use of the word "accurate" is her own, not grounded in the literature in math or statistics education. But rather than intervening as researchers familiar with the literature to refine how this term could have been better used or perhaps suggesting a different term that would be more appropriate, our intention was to see where Emma took the conversation next.

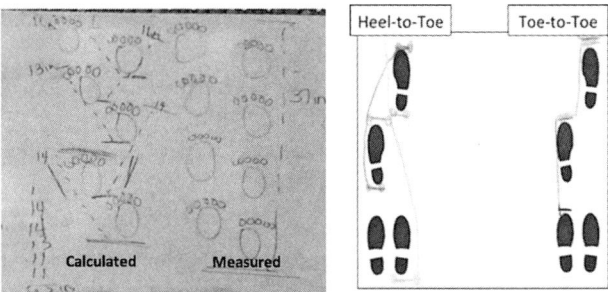

Figure 6.10
Refinement of step-size measurement convention from socially defined (heel-to-toe) to sociomathematically defined (toe-to-toe)

At this point, it is clear in the class discussion that the class agrees that step sizes are useful in knowing "how far you went," but given their newly discovered mismatch, they continued to disagree on where a step size should begin. This became clear as two students offered two different options: (1) measurement of the step size should begin at the heel of the first foot; and (2) measurement should begin at the toe of the first foot. Reminding the class of the question they were trying to answer ("how far someone went"), Emma asks the class to explain to her which one of the two "competing schools of thought" they "liked" and *why* they liked them. Martin explains to the class that he felt measurement should begin at the toe because "at the toe you *begin to go forward.*" Emma likes Martin's idea, and invites a student to join her in the front of the classroom to test Martin's assertion that forward movement begins at the toe. The student "raced" in front of the class, and Emma used their embodied actions as a context for reflection for the class, prompting them to think about where "movement begins and ends." This included leading the class discussion about how the weight of the body shifts to the toe before the foot finally leaves the ground to begin forward movement. The class also discussed what it would then mean to win the race, and they converged on the idea that typically the toe of the winner first crosses the finish line. The outcome of this disruption and the discussion that followed was the take-up of the "mathematically accurate" toe-to-toe measurement convention shown in Figure 6.10. This marked the genesis of a norm for what counts as a good measurement of step sizes, and the students then began to remeasure their step sizes using the newly accepted *convention.*

The connection between Emma's work here and Yackel and Cobb's notion of sociomathematical norms is obvious. Emma and her students succeed in reestablishing criteria for what counts as a "good measure" through deepening the mathematical grounding of the normatively accepted convention.[47] Emma's emphasis on *concreteness* and *accuracy* was critical in this effort, as was her pedagogical emphasis to represent continuous movement in the form of discrete steps. In terms of identifying the role of circulating references in Emma's work, it is noteworthy that Emma has not yet introduced ViMAP in the context of modeling motion using step sizes beyond the initial geometry unit. But, she is clearly inspired by ViMAP's affordance of representing continuous motion in a series of discrete steps that she had discovered in Vignette 1. As we mentioned in section 6.2, key to circulating references is the notion that different inscriptions and representational systems that highlight different aspects of the same phenomena are brought in contact with each other. Emmas's design of the activity reported in this vignette is such an example. It involved interpreting embodied movements and physical artifacts as data, generating both aggregate (total step size) and individual (the size of each step) measures, and a return to the embodied enactment and subsequent reflection in order to make explicit *why* the toe-to-toe measure is the more "accurate" measure. Furthermore, as the next episode will reveal, this activity is a precursor for modeling in ViMAP, as Emma shifts her focus to another sociomathematical norm: approximation.

6.6.3.2   Episode 2: Framing coding as "approximation"   In our weekly debriefs, Emma noted the difference between students' embodied experiences of walking and the ViMAP model of a Turtle moving forward in equal steps. Every step was different for each student, and Emma wanted students to notice this difference in the form of an activity so that the ViMAP model would become "concrete" (Emma's term) to the students. So, in the subsequent activities, Emma asked students to explore "approximation" and "prediction" (Emma's terms) as ways for them to refine the toe-to-toe measurement convention to "quickly" (Emma's term) measure distances that "could not be walked" (Emma's phrase) as well as a way of validating their ViMAP models of motion. To do so, Emma asked her students to consider how they might describe their stepsize data measured using the toe-to-toe convention with a single number, as opposed to reporting the size of each step as a different number.

Emma introduced the idea of "approximations" as values that are "kind of real" and "helpful." They are not exact measures of each step, but can be used to make predictions about the total distance traveled without having to measure each step separately. She framed thinking around approximate values as a measurement problem, asking the class to imagine a way to figure out "how far [someone] traveled after fifty or one hundred steps," but without actually walking and adding up all of the different step sizes.

Here, again, we briefly redescribe events that we have reported elsewhere, so that we can situate the relationship between sociomathematical norms and circulating references in Emma's classroom. Emma worked with the stepsize data from one of the students and asked the class to report what they noticed as she physically enacted the stepsize data collected by a student. Damien points out that as Emma was stepping forward, the size of each step was "changing." Emma agrees, but she follows up with another question: Is each step changing "by a lot" or "by a little?" She then asked the class to reexamine their own data with this question in mind. Damien responded that his steps "mostly change by a little," to which another student, Jayla, agrees. To create a contrast, Emma then humorously mimes walking with a dramatically changing gait per step. It is a walk where step sizes "change a lot" compared to the *approximate* step size. Laughing, students agree with Damien and Jayla's noticing that their steps were "about the same size."

Emma's next move involved engaging students in a ViMAP modeling activity to use their emerging understanding of approximate step sizes. She provided students with a hypothetical data set of step sizes: 11, 9, 11, and 12. She asked them to build a ViMAP model of the total distance traveled on the basis of the general pattern of the step sizes. To facilitate this, she asked students to reason about the following: if the hypothetical student "continued walking," what would "their next step be?" In a flurry of discussion, nearly all the students (14 out of 15) agreed that a "good" possible "future step" that was "close to the actual, but not exact." Many students also noted that a good value would be any of the values already given, because they represented the general trend of the values.

Emma then quickly introduced another activity with the specific goal to highlight multiplicative reasoning, by leading a class discussion on calculating approximate total distances when the size of each step was identical. She presented a hypothetical data set where a student moved forward by a step size of exactly 8 units over 6 steps, and asked students to use their lessons from math class on repeated addition and multiplication to calculate what the total distance would be. As students began to suggest that multiplying the number of steps with the step size might work, she explained that what the students were doing could be summarized in this way as a *formula* (*Total Distance = Number of Steps x Approximate Step Size*) which would allow the students to "find total distances that you can't actually walk," and therefore could be used to make "predictions."

Students were now beginning to see how they could use ViMAP as a tool for modeling their motion mathematically. An illustrative case is shown in the excerpt and the Figure that follow.

Excerpt: Angelo's prediction

Amanda: How far did you walk after taking 15 steps?
Angelo: 300 distance.
Amanda: That's exactly right.
Angelo: If somebody bet that I won't make it farther than 100 *I know* that I will make it.
Amanda: That's right. That's how a formula for approximate distance can help you, and if someone said "I bet Angelo would only walk 150 inches in 15 steps" . . . But you knew what your approximate step size was, could you prove them wrong?
Angelo: Yes
Amanda: How?
Angelo: I could look at my graph.
Amanda: Or you could do what?
Angelo: I could use a calculator. Fifteen times 20 equals 300.

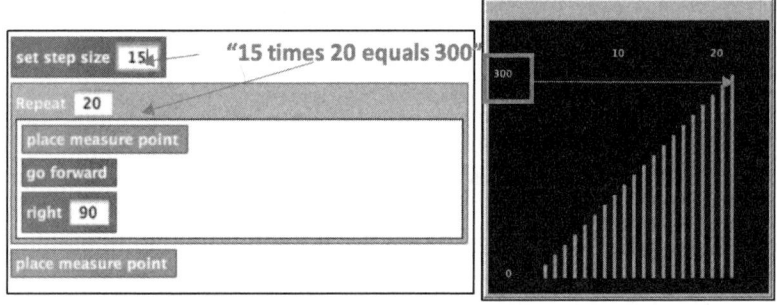

Figure 6.11
Angelo's ViMAP model

In this interview excerpt, Angelo (a student) interprets the multiplicative formula that Emma put up on the whiteboard as a means to both "win a bet" as well as mathematically verify the "accuracy" of his ViMAP model of distance using the graphs he had generated in his ViMAP model (shown in Figure 6.11) and his formula ("I could use a calculator"). Epistemologically, this is a significant move. As Angelo put it, using approximate values allows him to "know." We believe that Angelo's explanation of "betting" and "knowing" is his intu-

itive way of explaining what prediction is. Furthermore, this demonstrates that
Angelo is able to mathematically summarize discrete values to model continuous patterns of change.

Once again, as we saw previously, Emma established approximation as a sociomathematical norm by integrating multiple genres of representation ranging from embodied modeling to ViMAP modeling. She built her instructional moves in Episode 2 on the mathematical grounds that she established
in Episode 1. She emphasized modeling continuity using discrete measures
by drawing upon the same representational genres the class used in earlier
episodes. The push for concreteness was still present, but accuracy and approximation also emerged as powerful anchors in her work.

6.7   Growth in Computational Thinking in Year 1

A detailed and nuanced analysis of the growth in students' computational
thinking in Emma's class can be found elsewhere.[48] Note that although we
used the phrase computational thinking, we referred to facets of students' computational work as the primary data, and these facets comprise representational
practices central to design and scientific modeling. Yet, despite the focus on
practice, the sole reliance on computational work as the primary data might
imply that we were operating within a technocentric paradigm. We want to
highlight that the patterns of growth that we can infer from such analysis rest
on the much more intricate and rich observational work that we have reported
previously.

Here we report a general trend in Figure 6.12, which shows how students'
use of the ViMAP programming commands became increasingly sophisticated
throughout the duration of the activities reported here (Phase II). We scored
each student's final ViMAP model at the end of each class period in terms
of whether they used appropriate variables in their ViMAP code, and whether
their graphs represented appropriate element(s) of the phenomenon being simulated in their ViMAP code, each on a scale of 0–3. A score of 0 meant that
none of the variables used were appropriate, whereas a score of 3 meant that
there was no use of redundant or incorrect variables. The accuracy of the
graphs in students' later models was indicative of the appropriate use of the
"repeat" command and the order of placement of the "place measure" command. This in turn relied on a conceptual understanding of when to initialize
the measurement and how often the desired measurement had to be repeated
in order to generate the graph. In computer programming, *initialization* is the
assignment of an initial value for a data object or variable, and repeating the
measurement involved the use of *loops*.

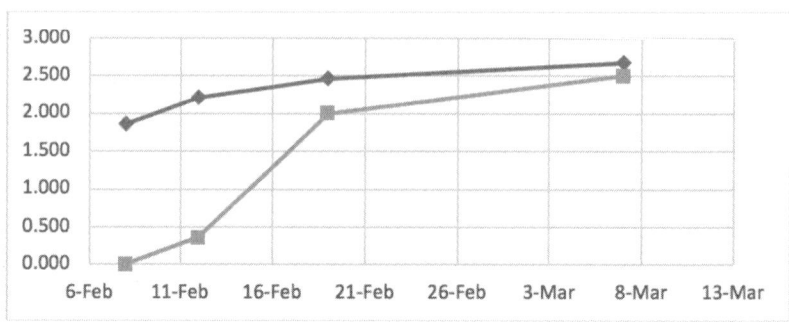

Figure 6.12
Improvement in computational thinking in terms of, *top line*, use of variables and, *bottom line*, loops and initialization

It is important to note that by itself, the graph shown in Figure 6.12 merely offers a device-centered view. That is, it restricts our understanding of the nature of the experiences to a radically reduced view of their device-level engagement. The illustrative cases we presented, on the contrary, illustrate how and why this improvement happened. They show that the development, deployment, and refinement of sociomathematical norms—central to which was repeatedly asking the question "what makes this code good from a *mathematical* perspective?"—was a key factor, and that working with ViMAP was deeply intertwined with other forms of representation and modeling outside the computer. Students, working with Emma, had developed an expectation that for their ViMAP code to be "good," it must do some mathematical work, and it also must represent their embodied experiences with a certain degree of fidelity (e.g, the total distance traveled by the ViMAP Turtle should be nearly equal to the total distance traveled by the students during the stamping activity). As "accuracy" became a "taken as shared"[49] disposition for students, coding became tied to iterative attempts to refine mathematical representations and reasoning, such as interpreting graphs and multiplicative reasoning.

6.8   Continuity, Heterogeneity, and Multivoicing as Emergent Tensions in Year 1

The story of Year 1 that we have described so far highlights how coding circulates across multiple forms of representations: ViMAP commands, embodied modeling, mathematical representations such as graphs and multiplication charts, and in the form of linguistic references that Emma invented. The framing of code as mathematical work—through the establishment of what appears to be similar to Cobb and his colleagues' formulation of sociomathematical norms—represents Emma's attempt to create an experience of *continuity* across these heterogeneous references.

We see this continuity as synthetic, that is, as an effort to bring together forms of activities, some of which are "concrete" and some of which are "abstract" (in Emma's words). Embodied activities made code concrete for students, because it enabled them to ground reasoning about computational abstractions in their own movement. Mathematical inscriptions also framed coding as transformations of representations that were already familiar to students (graphs and multiplication tables). In situating code among these experiences, Emma also positioned students as learners with some agency, giving them an opportunity to voice code as they are trying to describe and represent their movement mathematically. This emphasis on synthesis, one can easily argue, was designed by Emma in order to help her students and herself voice code, so that they could bring in familiar representations and discourse as legitimate experiences of coding.

And yet, the students' ViMAP programs show little representational diversity. Visually, they are all highly similar to one another, and all use STEP-SIZE as the Turtle variable to represent speed and motion. Furthermore, as Angelo's work and explanations in Vignette 3, Episode 2, show, students had begun to see the purpose of their work on ViMAP as the generation of graphs, which could then be used to "verify" if their code "accurately" represents their embodied experiences. The pitfall of such a framing of coding is that it can render redundant the representational palette offered in the programming language—there is no incentive for the students to explore different ways of computationally representing their embodied movement. Therefore, although synthesizing different representations of motion can be seen as a form of multivoicing, it does not necessarily encourage multivoicing from the perspective of computational expressivity.

The tension between univocality and multivoicing is inherent in *any* form of voicing due to the inherently heterogeneous nature of language, as Bakhtin and Todorov have argued (see section 2.1). We posit that the source of this tension, in our case, is Emma's emphasis on the *reversibility* of references—that is, being able to reversibly transform representational forms. Students' ViMAP code of motion eventually had to produce a graph of total distance traveled, and this value had to match students' embodied modeling experiences. In contrast, in Year 2, we will see a greater emphasis on understanding sources of error that prevent a complete reversibility between references as a reference is being transformed into another, and how this new emphasis on the experience of *transforming* references shapes computational expressivity in students' work.

6.9   Year 2: Amplifying Transformations across Circulating References

In Year 2, an important computational abstraction that circulated across students' physical, mathematical, and ViMAP models was the *measure flag*, represented in ViMAP by the PLACE-MEASURE-POINT command. Flags serve as indicators of the position of an object in time, and the idea of placing flags at the end of every "step" of motion was normatively accepted by the students as a useful representation for measurement of motion. The episodes we describe here illustrate how Emma highlighted the challenges emanating from measuring motion in this manner, and in doing so, how she broadened the scope of what coding in her science classroom could look like.

6.9.1   Vignette 1: Watching Videos, Noticing Error, and Modeling with Loops

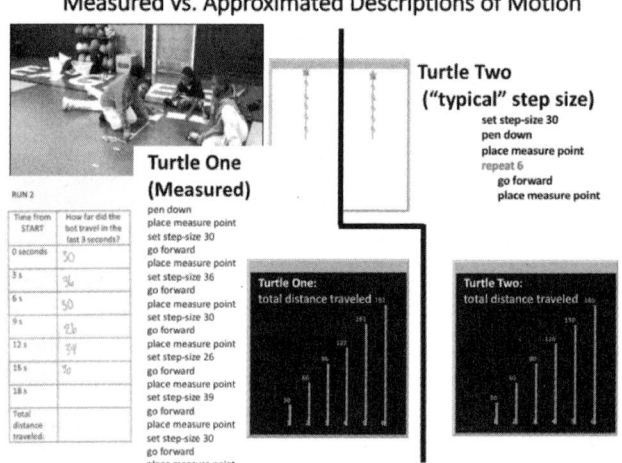

Figure 6.13
A student's ViMAP models of measured and "typical" step sizes

The work we are concerned with in this chapter from Year 2 began in October. Emma, along with the authors, designed this activity both as a refresher for her students who had continued with her from grade 3 (seven students), and to introduce the new students (14 students) to ViMAP in the context of learning science. The goal of the activity was for students to begin to describe motion in terms of distances traveled per unit of time and to use ViMAP to model this motion. Emma's objective was to help students "see" and describe the constancy of motion on their own. Emma continued to insist on integrating physical modeling with ViMAP. Along with Amy, she co-designed an activity in which students were provided with Lego Mindstorms NXT robots preprogrammed to move at a constant speed. The goal of this activity was for students to measure the distances traveled by the robot in regular intervals of time, so that they could describe the motion of the robot in terms of time-based units. Students were also provided with a measuring kit that included stopwatches, adhesive Post-it flags, and seamstress-style measuring tapes. Students coordinated the placement of position-marking flags with a stopwatch in order to come up with mathematical measures and explanations of robots moving at constant speed. Figure 6.13 shows a group of students measuring the distance between flags to find the distance traveled in each three-second interval. Students used these measurements to generate data, which were then used during the following two weeks to create computational models of the motion in ViMAP.

In their initial models, each group's measurements for the distances traveled in each three-second interval were nonuniform due to challenges inherent in the act of measurement, recording handwritten data, and transforming that data into ViMAP commands (see Figure 6.13). However, as we noted in earlier papers, none of the students problematized this issue by considering the limitations of the devices for measurement or the imperfect coordination of the placement of flags with the stopwatch.[50] The research team then designed an activity in which students watched a video of one group carrying out their measurement and data collection, and as a class, critiqued their work as shown in the video. Emma led this activity and replayed the video several times in order to support students to notice and to reflect upon successes and breakdowns in measurement. The students began to talk about error and inconsistencies in their measurement as resulting from *their* actions and not the robots' motion. For example, as Emma replayed segments of the video in which the group was trying to measure two successive "steps" of the robot, several students noticed that one of the measured steps was significantly longer than the others. In class discussions, students agreed that the timing of placement of the Post-it flag that demarcated adjacent steps was to be blamed, not the motion of the robot. In another segment, the video showed how the robot curved slightly as it moved. Several students were able to point this out while the video was playing and commented that curves did not show up in the way they made and recorded their measurement data. Another student also pointed out errors due to misreading the measuring tape, and in another case, due to the sticky flags' being unintentionally moved by getting stuck to students' shoes.

Emma then asked students to review their own measurements in order to determine what they believed was a "typical" distance measurement for their robot to travel in three seconds. Emma welcomed this as an opportunity to connect the modeling activity with learning about measures of central tendency, a topic that they were currently grappling with in their math classes. However, although the class discussion involved students identifying mean and median measures in their data sets, eventually each group settled on a number that could be used in their ViMAP programs as a loop, using the REPEAT command, in order to generate a value that is "close enough" to the total distance traveled by the robot. Prompted by Emma, students also began using the PLACE-MEASURE-POINT command, so that they could generate graphs from their Turtle graphics. Figure 6.13 summarizes the progression of this "constant speed" activity and compares one prototypical student's ViMAP programs for the measured step sizes and typical step sizes from their measured data sets.

6.9.2   Vignette 2: Flagging Videos and Modeling Friction

On December 2, students were working in small groups, designing investigations for measuring the acceleration of a marble rolling down an inclined plane. Emma framed the activity as follows: the students would have to design observations and collect measurements that would "convince [the principal] that the marble is changing speed." Students used physical analogues of ViMAP measure flags in the form of small adhesive page markers with arrow-shaped ends. Students' strategies for measurement varied in the number of measurements of step sizes during the journey of the marble. The following excerpt[51] of talk occurred as Emma questioned students to support their presentation of their evidence that the speed of the marble was increasing (Figure 6.14). In their material setup for sharing evidence, they positioned a tape measure alongside the ramp. They marked the midpoint of the ramp with a small adhesive page marker (measure flag) labeled END. In the following, bolded text is used to indicate when students point to part of the material setup as they explain it.

> Emma: Uh, real quick set-up question. Why are you guys measuring right now, just out of curiosity?
>
> Aden: To do the half-way point.
>
> Emma: Ooooo. I like that. How did you determine what the half-way point was?

Nylah: Because . . . we know that 2 divided into 48 is 24. So when we measure it we know that the . . . halfway point.

[Emma questions other groups about their measurement strategy, then refocuses attention on the presenting group.]

Emma: Okay. So first, you all tell me how you all are setting up your experiment . . . um . . . Aden, tell me a little bit and then I'll go to each one of you guys. What are you doing today for us today?

Aden: First, when we . . . the ball gets right **here** . . . [points to the midpoint "END" flag] Theo will start the timer, and then when it gets to the edge [points to end of 48" length] he will stop the timer. Then the next one, we are going to do one when [points to the beginning of the ramp] it gets to the **end** [points to the midpoint flag] when it starts from **here** [points to the beginning of the ramp] . . . he starts the timer and when it gets right **here** [midpoint flag] he stops it.

Emma: So you are going to take two different measurements of time, correct?

[Members of group nod.]

Emma: Mm-kay. Anything you want to add, Nylah? I mean what are you guys trying to prove? Are you trying to prove the same thing?

Nylah: We're trying to prove that the ball will accelerate or increase its speed.

Figure 6.14
Experimental setup for measurement of motion: small page-marker is placed at the midpoint of the ramp to indicate where to stop measuring

As the students collected and shared their measurements, Emma organized them in a diagram on the whiteboard (Figure 6.15). Her diagram shows the following: the students released the marble from the top of the ramp, started the timer as the marble reached the midpoint of the ramp, and reported the duration it took to travel from the midpoint to the bottom of the ramp. They measured this time as 1.4 seconds. Next, the students again released the marble from the top of the ramp, this time starting the timer at the release, and stopped the timer at the midpoint. They reported this time as 2.4 seconds. In her diagram, Emma labeled the beginning, midpoint, and end of the ramp flag at the midpoint as A (not yet labeled in Figure 6.14), B, and C. She asked the students to compare the two time intervals (1.4 s and 2.4 s), and led a class discussion about how her diagram could be construed as evidence of the marble's speeding up. It was not at all challenging for students to explain that in the second half of the motion, the marble was traveling faster than in the first half, as was clear in the discussion we recorded.

Figure 6.15
Emma helps to organize features of the students' investigation shown in Figure 6.14 on the whiteboard with a diagram.

Emma, however, wanted students to get to a more detailed description of discrete measurements of continuous motion. She noted to the researchers that the power of ViMAP was that it allowed students to measure "as many times as needed," whereas the physical world makes such measurements challenging. The ViMAP step size can be as small as it needs to be, whereas splitting the length of the ramp in half allows for only two measurements. To address this issue, Emma and Amy co-designed an activity in which students would video-record their physical measurements and then analyze them using the slow-motion replay functionality, which in turn would allow them to conduct finer grain measurements. The ViMAP measure-flags took on physical forms as Post-it arrows, where each flag represented a frame of the recorded video, as shown in Figure 6.16. This was a conceptually significant move, in which

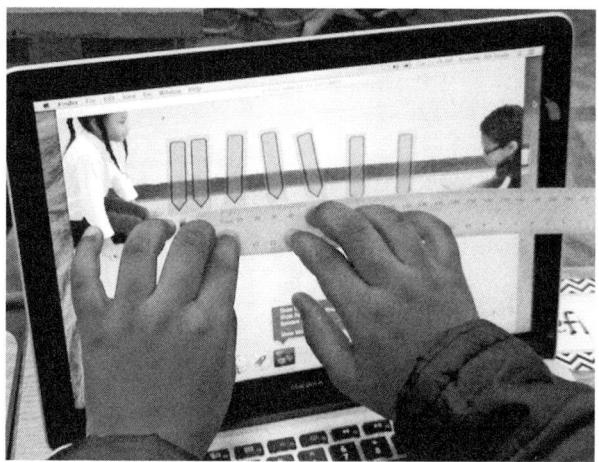

Figure 6.16
Flags as discrete measures: small adhesive flags mark the position of the marble at a 15-frame interval in the students' video of the marble's acceleration down an inclined plane.

students began to associate positions of the marble with intervals of time even more explicitly. In the friction unit (Figure 6.17), students made comparisons of total time traveled on the basis of these marked flags. This comparison created a contentious discussion: students were comparing the time in which a toy car slows down on two surfaces—on a rug (Figure 6.17, left) and on the tile floor (Figure 6.17, right).

Due to differences in how students placed the camera at the time of recording, students perceived the total distance traveled by the car on the rug to be *longer* than that of the car on the tile. They were looking at still images taken from video, and in these images, the total distance from (1) the flag that is farthest to the left and (2) the flag farthest to the right was a greater distance on the rug than on the tile. This did not match their experience—students contended that the car had gone farther on the tile than it had on the rug.

Figure 6.17
Setup and ViMAP model for measuring car path. *Left,* car path from launcher on a rug with measurements every 10 frames. *Right,* car path from launcher on tile hallway, measurements every 10 frames. *Bottom,* ViMAP model in which a student describes the car on the tile floor (squirrel) and the car on the rug (butterfly).

The evidence that eventually resolved this issue within class discussion came in the form of subtracting the time required for the motion to stop. The force applied to the cars was approximately equal, as students had used the same toy launch device on both the rug and the tile floor. In their distance measurements taken on top of the images, they had marked the flags with frame numbers (see Figure 6.17. Upon Amanda's suggestion that these frame numbers could be used as a form of evidence, students calculated the total time of motion on each surface: 118 frames–38 frames = 80 frames on hard tile floor, which they compared to 108 frames–38 frames = 70 frames on the rug. Although the difference of 10 frames is quite minor, the students were satisfied that they had produced evidence that it takes more time to slow down to a stop on the tile. Figure 6.17 also shows a student's ViMAP model of the comparison, in which computational measure flags (PLACE-MEASURE-POINT) are used to symbolize physical flags marking position in video-recorded motion, and the variable for decreasing step size (SET STEP-SIZE MINUS X) computationally represents the coefficient of friction on that surface.

6.10   Growth in Students' Computational Expressivity in Year 2

The growth in students' computational thinking in Year 2 can be better understood in terms of computational expressivity—that is, in terms of the range of their exploration of computational abstractions (variables) to represent motion. Here again we present an analysis similar to section 6.7. That is, we analyze students' computational work as primary data because it might help us see some of the effects of Emma's teaching on students' computational work. And again, similar to section 6.7, the point here is that we cannot simply make these connections between Emma's instruction and students' representational work by relying solely on technocentric analysis. Rich observations, like the ones reported in this chapter, help us understand how and why these changes may have occurred.

Initially, Emma's voicing was focused almost exclusively on using ViMAP to "verify" students' work outside the computer, similar to what we have seen in Year 1. She framed the use of ViMAP as a tool for generating graphs of distance traveled and speed. As a result, classroom conversations and student work with ViMAP both focused on using STEP-SIZE as the variable for representing speed in ViMAP. This was evident in their "constant speed" models of robot movement (October 23), as shown in Figure 6.18. Only one student used TURN and SHAPE as variables for representing speed, whereas another student (unsuccessfully) attempted to use commands for SECRET-NUMBER in his program.

With a greater emphasis on students' interpretive moves in physical modeling, we noticed that Emma also began to broaden the scope of computational representations of motion in students' work. In students' final models of constant acceleration (marbles moving down ramps) during the final week of January and first week of February, all students created programs that used variables other than STEP-SIZE to represent speed. Six students used RIGHT-TURN or LEFT-TURN, 16 students used the size of the pen (PEN-SIZE), 9 changed the agent size (TURTLE-SIZE), 12 changed the agent shape (SHAPE), 14 adjusted the heading (HEADING), and 3 used commands for secret number (SECRET-NUMBER). One student also adjusted the graphing settings to graph the color of the agent, rather than step size or total distance. All 19 students present in that part of the year continued to also use the ViMAP tools for measurement and graphs in some way in these models. The distribu-

tion of these different forms of representation in students' work is shown in
Figure 6.19.

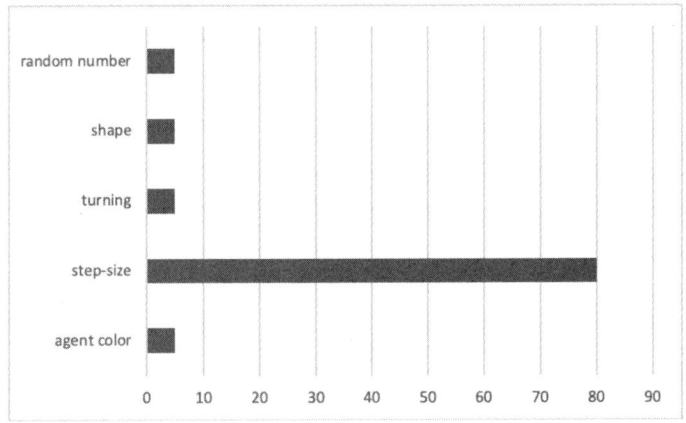

Figure 6.18
Exploration of variables for representation of speed in students' ViMAP models in October 2013

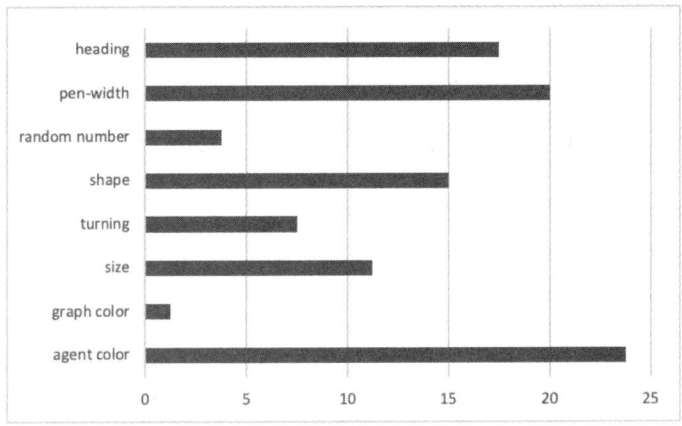

Figure 6.19
Variables for representation of speed in students' ViMAP models in early February 2014

6.11   Epilogue: Coding as Circulating Reference

Voicing code for Emma involved framing coding as circulating references in her classroom. Throughout Emma's journey, the persistent theme is the movement of the activities across forms of representational systems: embodied modeling, physical modeling, programming commands in ViMAP, and mathematical inscriptions that she would regularly use in her math lessons (e.g., multiplication tables, hand-drawn geometric shapes, and so on). As Emma began to *voice* code, *coding was the phenomenon that began circulating across these representations*.

At the heart of the notion of circulating references is the heterogeneity of representational genres. The defining characteristic of this heterogeneity is the transformation of one representational form into another, which implies a dialogical engagement between the forms. New meanings emerge as one form of representation is transformed into another: the ViMAP code becomes meaningful for both measurement of speed and geometric reasoning as it gets transformed into different forms of representations. Thus, it is through this experience of representational transformation that the conceptual field at play is amplified and disciplinary grounding is established, and certain elements of the conceptual field are brought to focus more prominently than others— often by revealing conceptual difficulties—which in turn deepen disciplinary engagement. For example, as Emma and her students began transforming their embodied movements to ViMAP programs through careful attention to mathematical thinking and inscriptions, they also began framing coding as an activity of mathematical modeling. Loops became mathematical tools that would focus students' attention to multiplicative reasoning, and the difficulties in transforming embodied movements to ViMAP commands focused students' attention to reasoning about angles.

So, in Emma's classroom, coding lives in translation between multiple representational forms. A line (or block) of code thus becomes a form of *intertext*:[52] an utterance that is meaningful only in relationship with other forms of physical, mathematical, and scientific utterances. It is transparent in that it stands in for something else; but it is also in dialogical relationship with the language it describes and analyzes and thus becomes a form of Bakhtinian *metalanguage*.[53] Early in Year 1, as Emma began to voice code, she also invented linguistic references to refer to phenomena that she encountered in ViMAP. As Vignettes 1 and 2 illustrated, regular polygons became "closed shapes" because the ViMAP Turtle had to return to its original position. Circles became shapes "with no straight edges" with ViMAP, in a bid to capture the tension between discrete steps that the Turtle must take in order to draw a "smooth" circle. Emma used the word "concrete" to indicate embodied actions corresponding to ViMAP commands for Turtle movement, and in Vignette 2, she even identified how certain programming commands can "hide" mathematical inconsistencies. In each of these cases, Emma was using natural language to delve deeper into computational abstractions. The phenomenon of coding, through Emma's (and her students') voicing, therefore circulated not only across different forms of inscriptions but also between natural and computer languages.

The invention of linguistic references continued throughout the two years that we worked with Emma. Later in Year 1, Emma introduced "accuracy" and "prediction" as qualities of measures that she hoped would pivot students' work from embodied and physical representations to computational models. But notable here is Emma's emphasis on *reversibility* of the transformations between these references: the embodied data had to be edited so that models could be "accurate" and "predictive," which in turn enabled students to use their ViMAP code and graphs to make realistic predictions about the total distance that they covered during their embodied activity. Emma also uses these opportunities to connect to the math curricular needs—the use of multiplication tables and drawing and reasoning about geometric shapes and properties of shapes. For Emma, ViMAP becomes a *lived-in* tool, one that she can use to teach math and motion, in a manner that she is already familiar with.

Emma's and her students' work reveals how coding can be reframed as reflexive with math and science in K–12 classrooms. However, there is an interesting shift in the way the heterogeneity in the circulating references is used pedagogically by Emma. In Year 1, too, she identified the mismatch between references. But in Year 2, especially in the vignettes we reported here, the pedagogical emphasis shifted to amplifying, rather than glossing over, the source of the mismatch. Emma designed reflective activities by bringing forward students' work as shared objects of reflection, and supporting classroom discourse around them. In terms of Bakhtinian perspectives on ambiguous and transparent discourse, such amplification relies on a greater accommodation from Emma toward the ambiguity in students' work—both in terms of the errors in their measurement and the interpretive dilemmas they experienced. The dialogicality of code became evident as Emma extended the modeling activities beyond the computer and the programming language. The PLACE-MEASURE-POINT command, which served as a useful but homogenizing functionality within the ViMAP world, became progressively more heterogeneous as it circulated across the ViMAP model, the physical act of measuring the marble's roll on the ramp, and the combined use of video frames and physical measurements. As the programming command became a circulating reference, it also supported dialogicality and heterogeneity by making explicit conflict among voices, as shared in Vignette 2. Students' computational work also showed evidence of greater diversity in terms of their use of computational abstractions. It is to this heterogeneous becoming of code that this chapter speaks, and it is through this experience of *bending* code toward dialogism, heterogeneity, and conflict among voices that Emma appropriated code and coding as a language for doing science in her classroom.

Notes

1. P. Sengupta, B. Brown, K. Rushton, & M. C. Shanahan (2018). Reframing coding as "mathematization" in the K–12 classroom: Views from teacher professional learning. *Alberta Science Education Journal*, 45(2), 28–36.
2. P. Sands, A. Yadav, & J. Good. (2018). Computational thinking in K–12: In-service teacher perceptions of computational thinking. In *Computational thinking in the STEM disciplines*, 151–164. Springer. See also: P. Sengupta, B. Kim, & M. C. Shanahan (2019). Playfully coding science: Views from preservice science teacher education. In *Critical, transdisciplinary and embodied approaches in STEM education*, 177–195. Springer.
3. L. McMahon (2017). Phenomenology as first-order perception: Speech, vision, and reflection in Merleau-Ponty. In *Perception and its development in Merleau-Ponty's phenomenology*, edited by K. Jacobson and J. Russon, 308–337. University of Toronto Press.
4. A. A. diSessa (2001). *Changing minds: Computers, learning, and literacy*. MIT Press.
5. See also: A. Vee (2013). Understanding computer programming as a literacy. *Literacy in Composition Studies*, 1(2), 42–64.
6. A. C. Dickes & A. V. Farris (2019). Beyond isolated competencies: Computational literacy in an elementary science classroom. In *Critical, transdisciplinary and embodied approaches in STEM education*, 131–149. Springer.
7. A. Pickering, ed. (1992). *Science as practice and culture*. University of Chicago Press.
8. R. Hall & J. G. Greeno (2008). Conceptual learning. In *21st century education: A reference handbook*, edited by T. Good, 212–221. Sage.
9. A. Pickering (1995). *The mangle of practice: Time, agency, and science*. University of Chicago Press.
10. B. Latour (1990). Drawing things together. In *Representation in scientific practice*, edited by M. Lynch and S. Woolgar, 19–68. MIT Press.
11. R. N. Giere (1988). *Explaining science: A cognitive approach*. University of Chicago Press.
12. R. Lehrer (2009). Designing to develop disciplinary dispositions: Modeling natural systems. *American Psychologist*, 64 (8), 759.
13. See chapter 1 for a detailed discussion.
14. B. Latour (1999). *Pandora's hope: Essays on the reality of science studies*. Harvard University Press.
15. Lehrer, Designing to develop, 2009.
16. A. A. diSessa (2004). Metarepresentation: Native competence and targets for instruction. *Cognition & Instruction, 22,* 293–331.
17. A. A. diSessa, D. Hammer, B. Sherin, & T. Kolpakowski (1991). Inventing graphing: Children's meta-representational expertise. *Journal of Mathematical Behavior*, 10(2), 117–160.
18. M. G. Ames (2019). *The charisma machine: The life, death, and legacy of One Laptop per Child*. MIT Press.
19. S. Papert (1980). *Mindstorms: Children, computers, and powerful ideas*. Basic Books.
20. Papert, *Mindstorms*, 1980, 32.
21. J. Margolis & A. Fisher (2002). *Unlocking the clubhouse: Women in computing*. MIT Press.

22. S. Vossoughi, P. K. Hooper, & M. Escudé (2016). Making through the lens of culture and power: Toward transformative visions for educational equity. *Harvard Education Review*, 86 (2), 216.

23. L. Martin (2015). The promise of the maker movement for education. *Journal of Pre-College Engineering Education Research* (J-PEER), 5(1), 30–39.

24. P. Sengupta, J. S. Kinnebrew, S. Basu, G. Biswas, & D. Clark (2013). Integrating computational thinking with K–12 science education using agent-based computation: A theoretical framework. *Education and Information Technologies*, 18(2), 351–380.

25. T. Todorov (1984). *Mikhail Bakhtin: The dialogical principle*. (Z. Goodzich, trans.) University of Minnesota Press.

26. P. Sengupta, B. Brown, K. Rushton, & M. C. Shanahan (2018). Reframing coding as "mathematization" in the K–12 classroom: Views from teacher professional learning. *Alberta Science Education Journal*, 45(2), 28–36.

27. See the following papers:

B. Sherin, A. A. diSessa, & D. Hammer (1993). Dynaturtle revisited: Learning physics through collaborative design of a computer model. *Interactive Learning Environments*, 3(2), 91–118.

A. V. Farris, A. C. Dickes, & P. Sengupta (2019). Learning to interpret measurement and motion in fourth grade computational modeling. *Science & Education*, 28(8), 927–956.

M. Guzdial (1994). Software-realized scaffolding to facilitate programming for science learning. *Interactive Learning Environments*, 4(1), 001–044.

P. Sengupta & A. V. Farris (2012, June). Learning kinematics in elementary grades using agent-based computational modeling: A visual programming-based approach. *Proceedings of the 11th International Conference on Interaction Design and Children*, 78–87. ACM.

28. For example, please refer to section 3.6.

29. J. Walkoe, M. Wilkerson, & A. Elby (2017). Technology-mediated teacher noticing: A goal for classroom practice, tool design, and professional development. *Proceedings of the 12th international conference on computer supported collaborative learning (CSCL 2017)*. International Society of the Learning Sciences.

30. R. Lehrer, L. Schauble, & D. Lucas (2008). Supporting development of the epistemology of inquiry. *Cognitive Development*, 23(4), 512–529.

31. M. J. Ford & E. A. Forman (2006). Redefining disciplinary learning in classroom contexts. *Review of Educational Research*, edited by J. Green & A. Luke, Vol. 30, 1–32. Washington, DC: American Educational Research Association.

R. Lehrer & L. Schauble (2000). Developing model-based reasoning in mathematics and science. *Journal of Applied Developmental Psychology*, 21(1), 39–48.

32. P. Cobb, T. Wood, E. Yackel, & B. McNeal (1992). Characteristics of classroom mathematics traditions: An interactional analysis. *American Educational Research Journal*, 29(3), 573–604.

P. Cobb, E. Yackel, & T. Wood (1989). Young children's emotional acts while doing mathematical problem solving. *Affect and mathematical problem solving: A new perspective*, edited by D. B. McLeod & V. M. Adams, 117–148. Springer.

R. Lehrer, L. Schauble, D. Strom, & M. Pligge (2001). Similarity of form and substance: Modeling material kind. *Cognition and Instruction: Twenty-five years of progress*, 39–74.

33. R. Lehrer & L. Schauble (2000). Developing model-based reasoning in mathematics and science. *Journal of Applied Developmental Psychology*, 21(1), 39–48.

34. A. C. Dickes, A. V. Farris, & P. Sengupta (2020). Sociomathematical norms for integrating coding and modeling with elementary science: A dialogical approach. *Journal of Science, Education and Technology*, 29(1).

35. K. McClain & P. Cobb (2001). An analysis of development of sociomathematical norms in one first-grade classroom. *Journal for Research in Mathematics Education*, 32(3), 236–266.

36. E. Yackel & P. Cobb (1996). Sociomathematical norms, argumentation, and autonomy in mathematics. *Journal for Research in Mathematics Education*, 22(4), 390–408.

E. Yackel, P. Cobb, & T. Wood (1991). Small-group interactions as a source of learning opportunities in second-grade mathematics. *Journal for Research in Mathematics Education*, 22(5), 390–408.

37. Yackel & Cobb, Sociomathematical norms, 1996, 470.

38. M. S. Gresalfi (2009). Taking up opportunities to learn: Constructing dispositions in mathematics classrooms. *The Journal of the Learning Sciences*, 18(3), 327–369.

39. Yackel & Cobb, Sociomathematical norms, 1996.

40. M. Lampert (1990). When the problem is not the question and the solution is not the answer: Mathematical knowing and teaching. *American Educational Research Journal*, 27, 29–63.

41. R. K. Yin (1994). *Case study research: Design and methods.* Sage.

42. B. Glaser & A. Strauss (1967). *The discovery of grounded theory: Strategies for qualitative research.* Weidenfeld & Nicholson.

43. Papert, *Mindstorms*, 1980.

44. J. Arnosky (2008). *Wild tracks!: A guide to nature's footprints.* Sterling Children's Books.

45. Yackel & Cobb, Sociomathematical norms, 1996.

46. Dickes, Farris, & Sengupta, Sociomathematical norms, 2020.

47. A more detailed discussion can be found in Dickes, Farris, & Sengupta (2020).

48. Dickes, Farris, & Sengupta, Sociomathematical norms, 2020.

49. Cobb, Wood, Yackel, & McNeal, Characteristics of classroom mathematics, 1992.

50. See the following:
Farris, Dickes, & Sengupta, Learning to interpret measurement, 2019.
A. V. Farris, A. C. Dickes, & P. Sengupta (2016). Development of disciplined interpretation using computational modeling in the elementary science classroom. In *Proceedings of the 12th International Conference of the Learning Sciences*, 282–289. International Society of the Learning Sciences.

51. See also: A. V. Farris, A. C. Dickes, & P. Sengupta (2020). Grounding computational abstractions in scientific experiences. In *The Interdisciplinarity of the Learning Sciences, 14th International Conference of the Learning Sciences (ICLS) 2020*, Volume 3, 1333–1340. International Society of the Learning Sciences.

52. Todorov, *Mikhail Bakhtin*, 1984.

53. Todorov, *Mikhail Bakhtin*, 1984, 23.

7 Coding as Aesthetic Experience

7.1 Aesthetic Experience and Views from the Margin

In this chapter, we reimagine coding as *aesthetic experience*,[1] a fundamental form of human experience that undergirds all other forms of experience and provides an essential continuity across heterogeneous forms of experience. Dewey argued that "the measure of the value of an experience lies in the perception of relationships or continuities to which it leads up."[2] He observed that the richness of an experience is marked by a variety of interests and an "intimate association" between various forms of knowledge, which are "torn asunder" and ruptured in schools due to the disconnected experiences of curricular domains. Positioning the problems of democracy as problems of education, Dewey located the discontinuity inherent in formal education at the heart of these problems. The possibilities and challenges of computing in K–12 education, we believe, to a large extent, are also rooted in a similar paradigm.[3]

Although Dewey initially located aesthetic experience in the work of artists,[4] he later argued that such experiences are fundamental in the context of scientific disciplines.[5] The fundamental nature of aesthetic experiences can be understood in the following senses: (1) they are predisciplinary and provide the substrate for any form of disciplinarily grounded sense-making; and (2) even in disciplinary contexts (including scientific disciplines), meaning is created through aesthetic experiences as we transform materials into expressive media. In the context of educational computing, the image of aesthetic experience necessarily brings into scope a much more complex repertoire of experiences beyond disciplinarily valued semiotic representations. In a previous paper,[6] we noted that this stands in stark contrast to how aesthetics in science and mathematics has been generally represented by the scientists and mathematicians themselves, as well as by STS scholars. For example, physicists' notions of the underlying "beauty" of their theories are premised on qualities such as coherence and interconnection.[7] The interconnectedness of facts becomes established through simplicity, symmetry, unity, and fundamentality of scientific representational and conceptual work, and these are aesthetic qualities that have been historically positioned at the heart of scientific and mathematical disciplinarity that are, by and large, reliable indicators for the empirical adequacy of theories.[8]

We have critiqued the studies of aesthetics in professional science as efforts to identify "beauty at the helm," as they focus on "the aesthetics of the interested—that is, accomplished scientists, who were deeply interested and thoroughly engaged in their professional pursuit."[9] The interconnectedness that makes Einstein's theories beautiful to Einstein's colleagues thus remains largely out of the reach of the broader public and young science learners. The heightened forms of engagement evident in the discussions of scientific and mathematical aesthetics also stand in contrast to the learning experiences of disempowered and the disinterested students who are typically left out of the fold of disciplinary engagement in classrooms. In the context of educational computing, the disciplinary *others* include women and gender minorities, racialized students, and students interested in the arts, most of whom do not identify themselves as competent in computing, even at the college level.[10]

How can such students, who are usually relegated to the margins of the classroom and the disciplines, begin to find their voices in computing in K–12 STEM? This is the question we seek to answer, and we believe that Dewey's notion of aesthetic experience, which also shares deep synergies with Bakhtin's positioning of aesthetics as *authoring*,[11] can offer us a valuable path forward. For Bakhtin, the activity of perception is the structure of authoring, thereby challenging the primacy of textuality over experience.[12] Warning against the "sterility and fixity of its formal manifestation,"[13] in moving beyond the printed symbol, Bakhtin argued that textuality is an "event" that involves acts of authoring or voicing of *utterances* which often brings together multiple "speech genres."[14] The reified forms that constitute text and the object of art, similar to our critique of the unconditional surrender to computational abstractions, are simply elements of the overall experience of art: the latter should not be subsumed by the former.

The transformative and synthetic qualities that Dewey noted to be constitutive of aesthetic experiences are also present in Bakhtin's notion of aesthetics. For text to become an aesthetic experience, it needs to be understood as heteroglossic. Bakhtin argued that elements of the text must be transformed into greater wholes that emerge from interpretive recombinations of these elements. It is through such transformations—for example, from text to "social voices"—

that language acquires a new "author," and enables the author to position both the self and the others in the horizon of experiences.[15] It is the difference and the *différance*[16] between the printed and the interpreted—between what is given to us and what can become of it—in which the window of aesthetic experience dwells. This is an invitation to "see more," rather than "merely seeing as."[17] It is an invitation to leave behind the lenses of instrumentality and device-level engagements and look beyond the semiotic subjugation of digital forms, in search of a truly expansive vision of coding as experience. To this end, we believe that the transformation of text to a social voice can be reimagined through Deweyan aesthetic experience that brings together disciplinary, personal and interpersonal, and historically grounded narratives and representations, while also calling into question curricular homogeneity and disciplinary masculinity, as we discuss in the next section.

7.2   Countering Pedagogical Homogeneity and Disciplinary Masculinity

Our goal here is to highlight that an emphasis on aesthetic experience, especially for students who have sociohistorically been positioned at the margins of the technoscientific disciplines, can bring to light how the masculinity inherent in these disciplines can be interrupted in educational computing classrooms. At the same time, it can also position students' lifeworlds outside the classroom as central to the learning experiences inside the classroom. This is not merely a matter of reorienting students' interests toward disciplinarily valued productions and performances, as that can still trap us within the technocentric trope, as we explained in chapter 2. Instead, what we highlight in this chapter is a far more complex issue, which interweaves curricular masculinity with experiences of marginalization in ways that are usually left out of the conversation in computing education.

To set the context, we return to one of the central problems we tackle in this book and have already introduced in the first two chapters: the emphasis on computational abstractions as the key element of computational work in the field of computing education. A careful look at the history of intelligence testing, for example, reminds us that the ability to use abstractions was regarded as evidence of intelligence that children of color were believed to lack. For example, the critical computing scholar Jane Margolis[18] pointed out that Lewis Terman, the intelligence-testing pioneer, wrote about immigrants, Mexican, and Black people:

> Their dullness seems to be racial . . . Children of this group should be segregated in special classes . . . [T]hey cannot master abstractions, but they can often be made efficient workers.[19]

Margolis further argued that while such views would not be voiced in these clearly racist terms today, the legacy persists. She wrote:

> The lasting effects of low expectations, a lack of access to rigorous courses, and de-facto tracking practices are that African American and Latino/a students are far more likely to be judged as having learning deficits, and to be placed disproportionately in low-track remedial programs where they have less access to high-status knowledge, powerful learning environments, and resources. (University of California Accord 2006)[20]

Margolis points out that most educational policy and curricular efforts to include computing in K–12 have been narrowly focused on technical solutions, for example, deployment of computers in inner-city schools, rather than also focusing on broader sociopolitical issues that shape students' and teachers' experiences. This technocentric approach often results in sustaining rather than countering deficit-based beliefs that teachers may hold toward racial minority students, and leads to the design of learning opportunities that do not provide students ample scope to learn about computing in a deep manner.

But the issue of exclusion is broader than the race of the learner. Historically, women have also been left out of computing, both in the classroom and at the workplace. We believe that the focus on the constructs of computational abstractions and thought processes as constructs that undergird the design of pedagogical activities, as well as the popularization of computational thinking, are forces for disciplinary homogeneity. This is evident in an overt focus on symbolic forms and reconfiguration of symbolic forms as pedagogical activities, which we posit reifies the masculinity of a "pure discipline." Such imaginations of computing, which primarily emphasize learning of representational forms that serve as disciplinary canons, leave out a vast scope of experience—both individual and shared—that renders these representations meaningful and, more importantly, can make the discipline personally and socially meaningful to the learners. An important part of this discourse takes the form of relational work, which also includes caring for others involved in the discourse,[21] friendship,[22] and emotion work.[23]

Our critique here is similar to Sandra Harding's critiques of the inherent masculinity of the discourse of a "pure mathematics." Harding argued that even the disciplinary value of mathematical axioms, which by definition do not need mathematical proof and cannot be proven mathematically, is determined socially and contextually as other mathematicians apply the axioms in different contexts to advance their mathematical proofs and arguments. The discourse of a pure mathematics, Harding argued, obfuscates (at the very least) or renders invisible the social and human lifeworld of symbolic mathematics in which mathematics comes alive. Such discourse renders invisible the relational work (which is regarded as feminine) that is actually at the core of creating meaning within disciplines.

Beyond the academy, masculinity in the high-tech workplace has similar effects of "disappearing" relational work and devaluing femininity, in which the language of a "pure discipline" gets transmogrified to policies and institutional cultures of "organizational effectiveness." For example, the gender and organization scholar Joyce Fletcher[24] conducted an ethnographic study of the construction of gender in a high-technology firm. By observing female design engineers in a high-tech firm, Fletcher illustrated that a striking feature of the organizational structure and culture in such institutions is "to disappear" or devalue relational practice, that is, activities that involve nurturing and community building work that are construed as "feminine." Fletcher notes that this devaluing does not mean that such work does not get done. To the contrary, women disproportionately continue to do such work and even face workplace harassment in the form of being name-called when they stop performing such emotion work. And yet, through the culture of masculinity reified in terms of the language of organizational effectiveness, there results an organizational blindness toward emotion work. Fletcher argued:

> The process of devaluing work that is associated with the feminine and reifying work associated with the masculine has probably produced many other routine but ineffective work practices—that is, practices that are in place not because they are particularly effective but because they are in line with masculine norms of behaving. . . . In theoretical terms, this links the issue of gender asymmetry (the way that devaluing the feminine is part of what it means to define oneself as masculine) with organizational effectiveness.

Fletcher positions her arguments from the standpoint of post-structuralist feminism, in which the very notion of experience is in peril. Post-structuralism views the production of knowledge as an exercise of power in which only some voices are heard and only some experience is counted as knowledge, thus bringing into question the notion of transcendent or universal truth. What is worth knowing, in this perspective, is ideologically determined, and Fletcher's point is that the ideological commitment to masculinity is what creates, reifies, and implements spaces where femininity—specifically, relational work—in the technological workplace is devalued and must be disappeared. Fletcher and Harding both termed such experiences, in which femininity is culturally and institutionally devalued and masculinity is simultaneously preferred, as forms of gender asymmetry. Fletcher further argued that the construction of such forms of gender-asymmetric experiences is also aided by language *in use*. For example, when women in Fletcher's study ceased performing relational and nurturing work, they were called derogatory names. At the same time, performance of relational work was not being recognized as "work" because of how workplace policies were laid out in terms of the language of "organizational effectiveness." That is, in a Whorfian sense, Fletcher illustrates how language in use also shapes experience, in this case, furthering a pattern of dominance of certain social realities that lead to the "disappearing" of femininity from the technological workplace.

Our position as critical phenomenologists draws upon post-structuralist feminist critiques of masculinity in technoscientific practice. This perspective critiques and challenges any account of "experience" that is presented without any account of how it is related to power. That is, rather than simply asking "what kind of experience is knowing," post-structuralists also ask "whose experiences are getting primacy in our accounts," and post-structural feminists in particular focus on bringing marginalized voices of women to the forefront. As phenomenologists, our commitment is to provide an account of the sense experiences of the members of the public who are at the center of code and coding in K–12 classrooms and yet are at the margins of the disciplines of computing and computational science. The post-structuralist feminist critiques of the masculinity inherent in pure disciplinarity in the hard sciences and mathematics and of the institutional masculinity in high-tech workplaces offer us lenses to understand the masculinity inherent in our (the authors') own conceptualizations of pedagogies of code and coding. This further helps us in seeing how in the cases reported in this chapter, the participants' actions and experiences can be seen as productive and powerful interruptions of masculinity.

As scholars who seek to advance the field of educational computing research, our goal here is to propose a *différant* language (for designers and researchers) that can help us think more carefully about the nature of experiences that can interrupt such forms of masculinity in pedagogical situations. By positioning computational science as "aesthetic experience," we orient our focus away from and beyond the language of "thinking" and "abstractions" that are hallmarks of "computational thinking." Critical aesthetics orient us to the richer forms of experience that result from interruptions of technocentric, individualist, and masculine formations in educational computing. The contrast between the terms *critical aesthetics* and *abstractions* is in invoking different forms of experiences as possibilities in the pedagogical space, which we hope can alter the relationship between power and knowledge in the context of learning to code.

And once again, we remind the reader that our position is not that computational abstractions do not matter, or that they do not exist. Rather, our goal here is to highlight that there are richer and more complex forms of experiences that cannot be subsumed by solely focusing on the production or use of computational abstractions, particularly in contexts of K–12 STEM education. By asking educational designers and researchers to consider alternative terminologies proposed here, our goal is to interrupt the pattern of dominance of "given" social realities—for example, viewing computational thinking as a "thought process," or viewing computational abstractions as the "key" to computational thinking—and propose new ways of conceptualizing and representing possible experiences of computing and coding, especially in the context of including marginalized voices.

7.3   Case 1: Shenice's Cross

Shenice (pseudonym) is a 5th-grade student in Ms. Ghanem's class. We worked in Ms. Ghanem's classroom twice a week for seven months to investigate how agent-based programming and modeling can be integrated in 5th-grade science and math. Ms. Ghanem taught in a 95% Black, public charter school in the mid-southern US. The school was established by Black social and religious activists with extensive support from the local Black community, who were concerned about the neglect of Black students in the post-segregation public education system in the US. This is a concern that is shared widely among Black parents in the US,[25] including families that sent their children to this school. Socioeconomically, students did not come from affluent families: 95% of the students received free meals in accordance with the National School Lunch Program. In terms of academic performance, standardized assessments for the two previous years indicated that Shenice's performance in reading, math, and science put her at risk for school failure.

The case we describe here centers around three computational models that Shenice created. Computationally, each of the subsequent models built on the previous one(s), even though each of the models represented a different phenomenon in the real world. Two connecting themes across these representations are (1) computational representations of continuous processes using discrete events of change and (2) Shenice's religious expression. We situate critical aesthetics at the intersection of these themes. This case illustrates how we have come to view Shenice's religiosity as central to understanding the deepening of her relationship with computing.

Along the first theme, Shenice's work typifies the affordances of Turtle geometry that children use to express emerging understandings about continuous change in science across our studies. However, the ViMAP models she built began also with a personally meaningful, cultural symbol—an *identity artifact*, which Leander[26] defined as "any instrument (sign, material object, embodied practice, etc.) that interactants make use of to shape the identity of an individual or group."[27] But it is not the charisma[28] of the computational identity artifact that we seek to identify; rather it is the unfolding of experiences leading up to and beyond the creation of the artifact that we delve into. This brings us to our second, deeply interconnected theme: Shenice's religiosity.

Like many students in the class, Shenice's local church played an important role in Shenice's life. This was evident in her conversations with friends in class about events that happened in church (some class members also went to the same church). She even invited some of the researchers, including Pratim, to visit her church. Despite the apparent disconnect between her religiosity and her computational modeling, this case illustrates the critical phenomenological possibilities (as well as missed opportunities) that live at the milieu of these worlds.

7.3.1 Shenice's First Cross

Shenice created her first ViMAP cross program during the first eight days of classroom work with us. During this period, our objective was for students to learn to use the programming commands to draw Turtle geometry shapes. The capstone activity for this first phase of learning was a quilt project, in which students individually designed squares of a computational quilt to be printed and reassembled into a physical paper quilt and displayed in the hallway. In future units, some students adapted some of their early Turtle geometry code from the quilts and reused it in models of scientific phenomena (e.g., motion), deepening their application of mathematical ideas and processes (e.g., multiplication) in these programs.

Shenice initially designed a simple cross for her contribution to the quilt project, which is shown in Figure 7.1. She asked for help from the authors to make her cross shape "more interesting." Shenice and Amy worked together trying various remixes of her code and Amy suggested she use the repeat command (that is, loops) and manipulate the turn angle. She wanted to make each side of the shape a different color, and once Amy showed her the programming command to do so, she also let the shape iterate many times, creating what she called a "star made of many crosses." She also programmed a second Turtle to draw a square in the middle of her star.

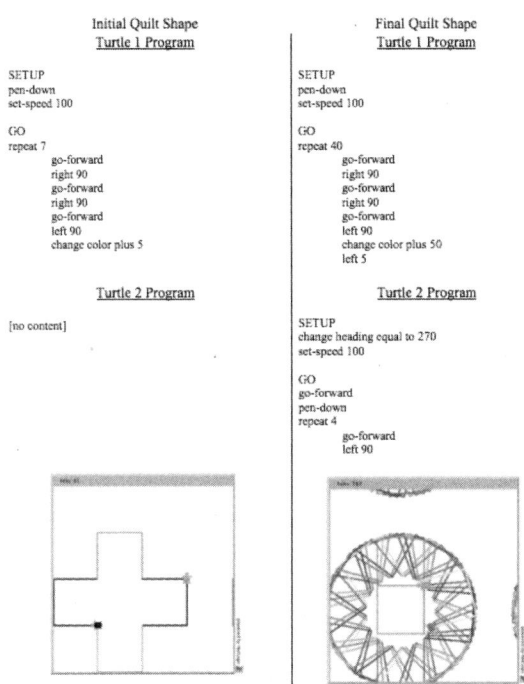

Figure 7.1
Shenice's initial and final quilt shapes during Phase 1

7.3.2   The Cross as a Model of Constant Speed

Four weeks later, on the 25th day of classroom work with us, students were asked to create a ViMAP program and graphs to model the movement of a robotic vacuum cleaner that traveled for four hours at a steady spread of 20 miles per hour. The goal was to encourage students to use their Turtle geometry shapes as models to represent motion. Screen capture data from Shenice's computer (see Figure 7.2) show that she went back to the cross that she had created a few weeks earlier and began modifying that program in order to represent constant speed for a duration of four hours by creating a square to indicate each hour of motion.

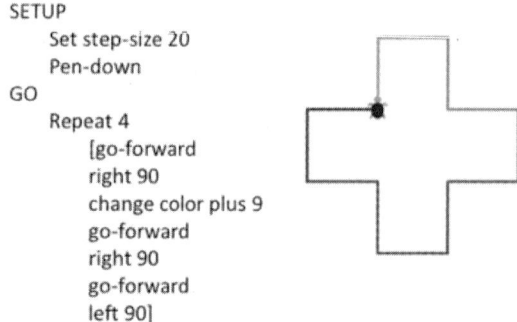

```
SETUP
    Set step-size 20
    Pen-down
GO
    Repeat 4
        [go-forward
        right 90
        change color plus 9
        go-forward
        right 90
        go-forward
        left 90]
```

Figure 7.2
Shenice's model of constant speed

She further modified her original program by changing the step size of the Turtle from 80 to 20 in order to represent the robot's speed of 20 mph. She used the command PLACE-MEASURE-POINT at the end of every step in her code to generate a speed-time graph and a distance-time graph. In her explanation of the model to the whole class, Shenice explained that each of the four squares of the shape represented how far the vacuum cleaner traveled in one hour. She also made a paper-based graph for speed and time showing four equal-height bars spread across an axis labeled "time." In Shenice's model, each of the four arms of the cross shape represented an equivalent unit of time of one hour.

7.3.3   Modeling Constant Acceleration

Two weeks later, students began to model a phenomenon we had been working with in the classroom: a ball rolling down a ramp for 12 seconds. Shenice continued to use the arms of the cross as the unit of time. In this model (see Figure 7.3), she completed the square so that each square represented one second. She also increased the side length of each square with each iteration of the square procedure. The program includes an initial (imaginary) square with a side length of 0. In order to represent constant acceleration, each square has a side length that is five units greater than the previous square and a pen width that is one unit larger than the previous square. The Turtle turns right by 25 degrees after drawing every square, thereby producing a spiral-like arrangement of increasingly larger squares. The associated graphs, also generated by the code in Shenice's model, is shown in Figure 7.4.

Shenice's model stood out in the class as no other student used a square as the unit of time. All students other than Shenice used the length of line segments (the ViMAP step size) to represent speed. In their models, if the Turtle was going faster, it would result in a longer step in the same amount of time. At the end of class, as the researchers were packing up, Pratim asked Shenice where she got the inspiration for her model, to which she replied that she was inspired by her "main man JC" (Jesus Christ). At this point, two of Shenice's classmates were also around her, and one of them asked Pratim which church he attended. Upon hearing Pratim's response that he did not go to a church (Pratim also explained that he is an atheist), all three expressed surprise through facial expressions, albeit playfully.

At this point, one of the students extended an invitation to Pratim to join her and Shenice at their church for service on the following Sunday morning. Pratim asked them what he would get to do if he went to church with them. Noting that there would be dancing and music, Shenice and her friends then began to dance around Pratim. Pratim asked Shenice if the trajectory of their dance resembled the squares in her ViMAP model. Shenice did not reply, and instead kept dancing and proclaiming "Hallelujah," with another student following her steps and joyous proclamations.

Figure 7.3
Shenice's model of constant acceleration

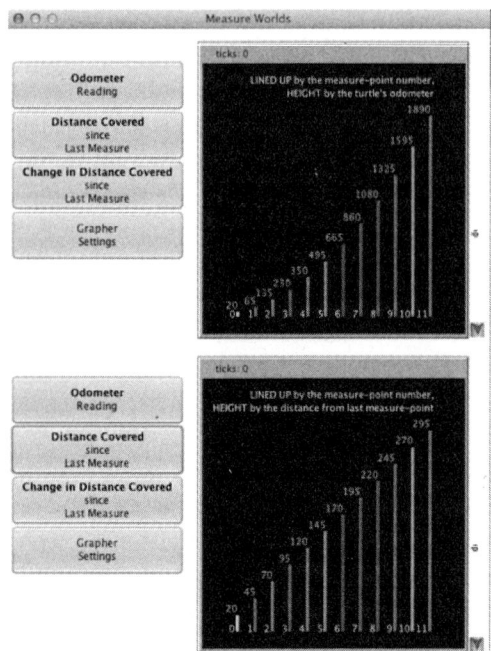

Figure 7.4
Shenice's graphs of constant acceleration

7.3.4   Analytic Summary

So, what makes this case an example of aesthetic experience? The most obvious answer to this question is that Shenice was able to integrate a personally meaningful religious symbol with the computational modeling work. In
viewing aesthetics from a Bakhtinian perspective, this case is also an example of *authoring*, through which ViMAP programming commands, similar to
Bakhtin's notion of *text* and textuality, *become* a "social voice." This image
of transformation of *text* to *voice* is akin to Bakhtin's analysis of works of art
as artistic *utterances* that result from dialogically bringing together multiple
speech genres (e.g., everyday speech and specialized text).

One could also reasonably point out here that Shenice's choice to reuse the
cross as a starting point for her model of constant acceleration could have been
due to other reasons besides her religiosity. For example, the usefulness of the
cross as a model of speed could have been a motivation, or it could have been
an idiosyncratic choice. But if we focused solely on the possibility of reuse and
redesign—that is, in the "positing of practice as driven by a rational orientation

toward the future"[29]—we could limit our understanding to how the students developed a specialized language that describes patterns of representation that are valued disciplinarily and also meet goals of self-fashioning. Therefore, despite evidence of students' agency, this would still remain an account of disciplinary authenticity, further reifying a device-centered image of the child entering the world of code through personal meaningfulness.

Instead of looking at Shenice's case as a story of *available designs* and *re-design*, we can position the story as a complex unfolding of experience, the heterogeneity in which might be invisible to us through reductive second-order perceptions such as commonly held views of computational thinking. Second-order perceptions are powerful enforcers of homogeneity. In focusing overtly on the progressive sophistication of students' computational work as the de-sired forms of experience, we can, unintentionally, become and remain un-aware of the rich possibilities in heterogeneous discourse. In this case, these possibilities might have included opportunities for us to engage more deeply with students' families and communities, which is particularly important given historical and continual forms of systemic marginalization and oppression of Black people in the US, and their exclusion from our figured worlds of techno-science.

It is also noteworthy (as we also have mentioned in chapter 1) that the his-tory of public education for Black Americans in the US is deeply tied to Black churches, as evident during the Freedom Schools movement, particularly in the Southern US.[30] The involvement of community members and churches were also prominent in Shenice's school, and ignoring Shenice and her friends' in-vitation to visit their Sunday service was, in hindsight, a missed opportunity to explore and deepen these connections. In this context, it is important to note that learning scientists have shown that deepening connection with marginal-ized students' communities can not only center their lived experiences within the classroom, but also deepen disciplinary complexity in STEM education.[31]

What could become of code and science in such heterogeneous spaces is not merely a matter of future-making—but also a renewal of community and history—opportunities that were not recognized (and therefore, not taken up) by the researchers. If taken up, however, it is likely that such acts of différance would render code and science strange, perhaps beyond immediate recogni-tion through the usual lenses of computational thinking and science education. This is at once an epistemological, axiological, and methodological moment, and we must be changed in these moments, instead of rendering them into homogeneous forms that are recognizable only through the hegemonic, uni-versalist lenses of technoscience. The push for homogeneity was evident in

Pratim's attempt to link the dancing and singing to Shenice's computational work, which Shenice fittingly ignored.

Critical phenomenology then, is the call for us to be alive to these moments of emergence that are also critically and historically grounded. A critical phenomenological commitment should serve as a continuous reminder that might prevent such moments from being rendered illegitimate as we focus on accounting for computational abstractions in students' experiences. Voicing as aesthetic experience brings into account these dimensions of human experience that remain untouched in most assessments of computational thinking. It prepares us for the emergence of a liminal space where accounts of world-making merge with a sense of "belonging-togetherness that takes ontological precedence over discreteness of components and subject-object separation."[32] Such a space can, unexpectedly, bring together Shenice's fellow churchgoers in the classroom, her favorite activities, (singing and dancing), and coding the physics of motion.

It is in making these connections visible that her ViMAP model of the cross becomes an example of Bakhtinian transparent discourse. And, even though as researchers, we failed to recognize the extent of richness that such transparencies can offer in terms of connecting with students' communities, this is still an image of heterogeneity—of many voices coming together and a rupture of curricular and disciplinary homogeneity beyond code, through performances that model caring and friendship, not merely the physics involved. A critical aesthetics of code, then, should also remind us that learning to code must not be devoid of caring for each other, and we can begin to see how voices from the margins can re-position their work at the center of the discipline through valuing *relational work*. We explore this theme further in the following illustrative case.

Finally, it is worth spending a few moments thinking about what we stand to lose when we assess Shenice's work by stripping off the richness of her voicing. An example is shown in the graph in Figure 7.5. It shows an analysis of Shenice's computational work, including the work reported here, in terms of different facets of computational thinking that we identified on the basis of the commonly used measures reported in the literature (e.g., appropriate use of computational abstractions such as variables and loops, and scientific modeling practices such as iterative design). The graph is rather unassuming; it positions Shenice's scores alongside the class average. One could argue that this graph helps us position Shenice's work in light of her ability to use the institutionally and disciplinarily valued representations, which include computational abstractions (loops and variables), design thinking (iterative refinement of her ViMAP program), and mathematical reasoning (reasoning about angles, shapes, and multiplication). But it is in the phenomenological account reported here that Shenice's voice comes to light. This account certainly can include analysis of her computational work, but should never be subsumed by it.

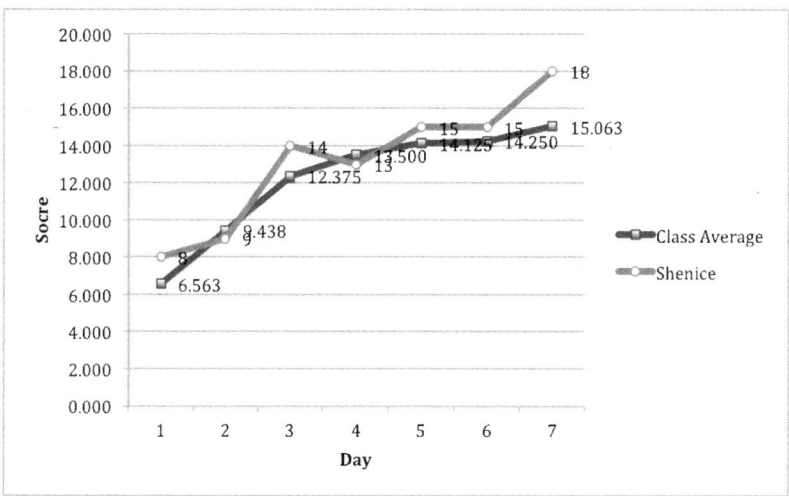

Figure 7.5
Shenice's models scored on the basis of her use of computational abstractions, design thinking, and mathematical reasoning during the first seven days (spread over four weeks) of the class

7.4   Case 2: Ariana and Matt's "Thomas"

7.4.1   Setting the Stage

Matt and Ariana are two 5th-grade students enrolled in a two-week summer course on agent-based computer modeling for learning Newtonian mechanics that we (the authors) co-taught on a university campus several years back. There were 14 students in the course, admitted on a first-come-first-served basis. We had taught four similar out-of-school courses in previous years. Usually these courses offered us opportunities to pilot different versions of ViMAP and learning activities. At this time (mid-2012), coding camps and robotics camps for kids had not yet become as popular as they are now, even though there was a high level of interest among local schools and parents to send their children to our courses. As a result, admission in our summer courses was highly sought after, reflected in the high number of applications compared to the number of available spots.

During the first couple of days in the course, Ariana and Matt each disclosed to the authors that they had no interest in computer programming. Ariana was especially interested in history and literature, and Matt was an aspiring actor. Neither saw themselves as people who might enjoy or be good at computer programming, and both of them had joined the course because of their parents' insistence.

From our perspective, as instructors of the course, the central disciplinary learning goals for students in terms of learning programming and physics were (1) to develop fluency with agent-based programming and modeling motion as a process of continuous change; and (2) in the process, begin to develop deep conceptual understandings of the relationships among distance, speed, and acceleration.

During the first day, Ariana completed the introductory activity of drawing some simple Logo-based shapes such as squares, rectangles, circles, and triangles. Our earlier studies had shown that these activities are helpful in familiarizing students with the programming commands as well as computational abstractions such as variables and loops. Furthermore, later in the course, students are introduced to different phenomena that involve constant speed and constant acceleration, and the same mathematical shapes can also serve as models of these phenomena.[33] So, in our perspectives, engaging in these shape-drawing activities was important for curricular flow, as well as conceptual development of the learners.

The second activity, immediately following the shape-drawing activity, was designed to ramp up the complexity involved in programming, while at the same time creating a context that would be personally meaningful for the learner. It involved students' creating computer programs to draw one letter from their own names. The class was allotted an hour for their work on this activity, and we noticed that neither Matt nor Ariana was really willing to move on to other activities. Matt had not drawn any letter of his own; instead, we found him working with Ariana to help her create a program for writing "THOMAS." Matt and Ariana spent the remaining part of the day working on this project, and they continued this work over the next three days as a side project, while (and sometimes instead of) working on other assigned projects.

7.4.2   Care and Humor in Drawing "Thomas"

It took us three days to realize the extent of Matt and Ariana's engagement with the THOMAS project. On the third day, while the rest of the class was working on modeling motion of a roller coaster using ViMAP and TuneBlocks,

Matt and Ariana continued to work together on programming ViMAP to draw THOMAS. Surprised at their persistence, Pratim asked Ariana why she was so interested in writing out THOMAS, instead of writing only one of the letters of her own name as the rest of her classmates did. The conversation that ensued is reported in Figure 7.6.

Ariana explained that the name *Thomas* was special for her because her best friend's name is Thomas, with whom she is "really, really, really" close. But there are also three other Thomases in her life: "The fourth Thomas is my best friend" [line 14, Figure 7.6], "the third Thomas is from *The Maze Runner*" [line 16, Figure 7.6], and "the second Thomas is the maze runner's dad, and the first Thomas is Thomas Edison, the scientist. And this will be our fifth Thomas, our little Turtle here. And he is so good. He is going to preschool and he is knowing how to spell his name" [lines 18–21, Figure 7.6].

As we have argued elsewhere,[34] this is an example of how the ViMAP Turtle became a transitional object in Ariana's work. It exists as a computational object that is symbolically constructed, and yet at the same time, it also represents her fondness for her best friend who is her neighbor and classmate and "really, really close" to her, and her favorite fictional character from a young adult book series (Thomas in *The Maze Runner*). The ViMAP Turtle, as Ariana points out in her interview, is the fifth and youngest Thomas in her life [line 20]. Ariana positions Turtle-Thomas as a preschooler who is learning how to spell his name, clearly indicating that the Turtle has also become an object of affection for her.

Matt's reactions to Ariana's story of Thomases is also notable. It is clear that Matt has been subjected to prior conversations about Thomases, and as he mentioned to Pratim later, "that is all Ariana would talk about." So, when Pratim asked Ariana to explain why she chose the name Thomas, Matt immediately started saying "don't go through the list! don't go through the list!," and further in the conversation "it burns! It buu-ur-urns!" Throughout the interview, Matt continues to playfully pretend, making faces, with a hint of a smile, how this conversation is unbearable for him. In line 24, for example, he says "heeeelp me . . . help me!" with hands extended to the camera. Ariana continued with her responses, ignoring Matt's playful interruptions, and it became clear to Pratim (the interviewer) that Matt and Ariana had by now established a relationship where there was space for sharing personal narratives and humor.

1.	ARIANA:	Well, my best friend's Thomas and I really, really, really, we're
2.		really, really close..see, he's like my neighbor [and he's also my
3.		classmate
4.	MATT:	don't go through the list!] don't go through the list
5.	ARIANA:	okay=
6.	MATT:	=too many Thomases, way too many Thomases
7.	ARIANA:	and, [oh yeah, I made a video about how..
8.	MATT:	*(whispering to the camera)* too many, too many]
9.	ARIANA:	how. I made a video about an [explanation of the Thomases
10.	MATT:	*(moaning)* I said don't go through with it!]
11.	ARIANA:	so it's Thomas *(counting on fingers)* named after Thomas who's
12.		named after Thomas, who's named after Thomas
13.	MATT:	*(covering ears)* it buuuu[urrrr-urrrr-urrrr-urrrrns!!
14.	ARIANA:	The fourth Thomas is my best friend]
15.	MATT:	It burns! It bur-ur-urns!
16.	ARIANA:	the third Thomas is..um..from *The Maze Runner*
17.	MATT:	*(whispering)* It burns!
18.	ARIANA:	and the second Thomas is the maze runner's..dad...and the first
19.		Thomas is Thomas Edison..the scientist. And this will be our fifth
20.		Thomas, our little Turtle here and he is so good..he is going to
21.		preschool and he is knowing how to [spell his name
22.	TEACHER:	So the turtle is going to preschool and he is knowing how to [spell
23.		his name?]
24.	MATT:	[heeelp me, help me! (hands extended to the camera)]
25.	ARIANA:	Yes, basically...
26.	TEACHER:	All right.

Figure 7.6
Ariana and Matt's interview transcript

This playful banter was not simply a personal space for Ariana and Matt: Ariana also told Pratim that Matt has been helping her create the program by planning the space allowed for each letter within the simulation. In this early version of ViMAP, students used PEN-UP, FORWARD, PEN-DOWN, and the LEFT-TURN and RIGHT-TURN commands to reposition the Turtle for each letter. Matt used the forward command to calculate the maximum width of the enactment space in ViMAP, and he partitioned that width into space allotments for each letter of the word *Thomas*, estimated the vertical adjustments needed to reposition the Turtle after it completed drawing a letter, and recorded those in his notebook. Ariana then used these plans to determine the spacing for the letters.

By partitioning the geometric space of the ViMAP simulation world, Matt (and Ariana) had invented their own version of Cartesian coordinates for the beginning and end of each letter. On the basis of Matt and Ariana's work, we decided to implement a new programming command in the ViMAP library of commands—SETXY [X-Position][Y-Position]—which allows users to specify the location of the Turtle based on its x and y coordinates.

7.4.3 Analytic Summary

Similarly to Shenice's case, the connections to Deweyan aesthetic experience are obvious. In terms of Dewey's concerns about bringing together disparate disciplinary domains of knowledge (the synthetic dimension of aesthetic experience), it is evident that creating programs to write "THOMAS" brought together the domains of computing and geometry, alongside Ariana's personal narratives and Matt's humor. Programming involved the successful use of relevant computational abstractions, such as variables and loops, and simulating the trajectory of the ViMAP Turtle in the shape of each letter involved significant complexity in terms of figuring out position of the pixels at the beginning of each letter. In contrast, the instructor-mandated activity of drawing only a single alphabet letter would have involved a far less extensive exploration of both mathematical reasoning and programming. The transformational quality of aesthetic experience is also evident: not only did the students transform the goal of the activity into personally meaningful and collaborative work, but they also were able to create computational objects that were mathematically meaningful.

Also notable here is the emergence of *carework*, which is a dimension of experience that is difficult to capture using only the lens of computational abstractions. This is evident in several places in the story we shared. Ariana's positioning of the Turtle as the "youngest Thomas" who goes to preschool and therefore must learn the alphabet by writing his own name, is an example of how learning to program was also about learning to care for the ViMAP Turtle, which had now become the fifth Thomas in her life. Matt, who desired to be a comedic actor, creates a space for humor through playful expressions of aggravation ("it buuuurns") with Ariana's repeated narrations of her relationship with the Thomases in her life, but has deserted his own work in service of being Ariana's personal geometry assistant. The comedic moment is also unplanned, as it emerged when Pratim, unaware of the history of Thomases, asked Ariana to explain why she chose to write out "Thomas." Inventing their own version of a Cartesian representation of the ViMAP simulation world was no small feat, and Matt took on the bulk of this work himself while Ariana worked on the letters. The ViMAP Turtle on Ariana's screen thus became another example of Bakhtin's transparent discourse, not only standing in for disciplinary domains, but also creating a liminal space for caring and friendship between

Ariana and Matt, who continued to rupture curricular homogeneity throughout the remainder of the course.[35]

This brings us back to Joyce Fletcher's critique of cultures of masculinity that reside in technological spaces and are also encouraged institutionally.[36] Fletcher positions carework as an antithesis of institutional policies of efficiency and effectiveness that are held as measures of technological prowess and success. Fletcher illustrates how caring for co-workers, in the form of helping co-workers in challenging and complex technology design projects (similar to Ariana and Matt's work on drawing *THOMAS*), and in helping co-workers navigate complex interpersonal issues at the workplace, are regarded as "feminine" behaviors and are institutionally devalued. The effect of such institutional doctrines and cultural expectations of masculinity is to make such work "disappear," thereby acting as deterrents for women in technological spaces, who typically shoulder the bulk of such work. Ariana and Matt's case can therefore serve as an important example where relational work was central to their learning, rupturing curricular homogeneity and disciplinary masculinities inherent in our instructional design, which initially did not create any space for such work.

7.5   Epilogue: A Critical Aesthetics of Coding

This chapter, like many of the other chapters in this book, offers an opportunity for a phenomenologically grounded *radical reflection*[37] on what form of experience coding can become, beyond the narrow focus on mastery of computational abstractions. In Merleau-Ponty's terms, it is through radical reflection that phenomenologists seek to give voice to the texture of experience in the very process of its happening by rupturing our familiarity with the phenomenal field. This rupture brings about new possibilities in "rendering strange"[38] what is already known and already seen. Here again, we see a deep synergy with Deweyan accounts of aesthetic experience, which as Higgins argued, calls for "seeing more" rather than merely "seeing as."[39] In aesthetic experience lies the possibility of challenging the dogmas, ideologies, and cultural expectations that stem from instrumentalism, because Higgins argues that "it is instrumentalism that has led to the marginalization of the aesthetic in education."[40]

To repeat an often-mentioned argument in this book, instrumentalist narratives and ideologies are still what dominate the discourse in computing[41] and, even more broadly, in design and innovation.[42] Positioning instrumentalism as a worldview that encompasses and advances a vision for public education in terms of the lowest common denominator of ethical life, Higgins wrote that instrumentalist ideas are edified in common parlance, such as "getting a living," "real world experience," etc. This rhetoric predefines living in terms of working at a job, and reality in terms of simple and taken-for-granted facts that we must face.[43]

But how can a focus on aesthetic experience really challenge these instrumental ideologies without creating what Higgins termed "aesthetic clichés?" Are we simply reiterating a phrase that lacks an experiential account and is bereft of possibilities for future expansive imaginations and actions by others? This is where the critical phenomenological agenda can be helpful, given its goal to bring into account new forms of *sense experiences from the margins*. In a truly critical sense of "seeing more," critical phenomenology can help us "see" and value accounts that have largely remained outside the scope of our considerations in the context of educational computing.

This chapter specifically argues that in order to bring such forms of experiences into account, we must work to bring to light potential places of rupture with the familiar ways of looking at coding that challenge the notions of disciplinary purity and masculinities. This is a form of radical reflection that is fundamental to phenomenology[44] and also relies on historical and institutional critiques of the contexts in which coding is taking place. It brings to light— and questions—*who* is(are) coding, and offers an axiological re-examination of what should count as coding. Our work is prompted by several alarming notes in the literature, as we have mentioned previously in the book. This includes the clearly racist history of how the term "abstractions" has been used by Terman, one of the pioneers of intelligence testing, as early as 1923,[45] and Margolis's poignant observation of how historically marginalized students are still left out of the fold of meaningful computing education in K–12.[46]

A critical aesthetics of coding is a direct response to important challenges that educational computing must overcome: the commitment to disciplinary authenticity in a narrow form (e.g., emphasizing reproduction of computational abstractions in curricular contexts) and the institutional emphasis on efficacy and efficiency (in both curricular and institutional contexts) that often results in "acts of disappearing."[47] What gets disappeared through such performances and expectations of homogeneity and masculinity are relational work and the ethic of togetherness, and these are precisely the places where we rupture with the familiar. The story of "seeing more" is thus also premised on identifying how our current ways of seeing lead to merely "seeing as," or worse—seeing less, or not seeing. Being aware of these dangers involves closely examining technological cultures and spaces through these lenses of disciplinary masculinity and computational homogeneity.

Finally, it is important to consider the following question: what do our accounts here add beyond Papert's notion of personal meaningfulness in the context of children's computational work? Central to Papert's vision of personal meaningfulness was the notion of a near-infinite malleability of the computational object in taking on the myriad shapes of children's imaginations, which also leads to the child's developing agency and control over the computer. This is a form of a machinic commitment that limits our vision of coding to *device-level* engagements. Once again, we direct the reader to Ames's poignant critique of the dissonance between the technocentric and market-driven imaginations underlying the design of One Laptop Per Child (OLPC) devices and how the devices were actually used by the intended users (children) in developing countries.[48] For example, Ames found that many children in Paraguay did not use the OLPC laptops not only because they were too difficult to use, but also because of the "fullness of their lives without the computer."[49]

What individualistic and device-centered accounts of personal meaningfulness generally lack are an emphasis on the discourse and performances that unfold outside and even away from the computer. In this technocentric imaginary, the malleability of the *protean* computer (Ames poignantly termed OLPC laptops "the Charisma machine")[50] stand in for the lived worlds of children. Although Papert did bring up the importance of paying attention to cultural and institutional contexts of learning and technology later in his critique of technocentrism, scholarship in the constructionist tradition has largely been tethered to the notion of device-level engagement. For example, even when researchers pay attention to breaking the disciplinary masculinity of coding by incorporating stitching and weaving in the form of e-textiles, there remains a looming danger of folding experiences outside the computer onto computational devices.[51] In contrast, in learning to see coding as aesthetic experience from a critical phenomenological lens, we intentionally and necessarily bring into account the lifeworlds of the learners, including *who* (not only *what*) they care about, and even their religious lives—*but* without folding these experiences onto computational devices. This allows us to trace the historical becoming of the technological creations of the learners in ways that extend far beyond narrow views of disciplinary authenticity grounded in the epistemology of control. The critical phenomenology of resisting instrumentalism and the cliché of personal meaningfulness necessitates that we become *alive* to relationality and account for the "fullness of their lives"[52] outside the computer.

Relationality, as Massumi argued, is fundamentally emergent. It cannot be accounted for simply in terms of the objective properties of the "discrete elements" in play or the interactions that might logically be predicted according to those properties.[53] The emergence of unexpected continuities between Shenice's church and her ViMAP model, the carework, the friendships—both old (in Shenice's case) and new (in Matt and Ariana's case)—are all examples of such relationality. The reductionist views of child-computer dyads, personal interests, and meaningfulness, are essentially committed to device-level engagements, and simply fall short of accounting for the richness inherent in relationality. Heterogeneity and emergence are *fundamental* to accounts of rationality from a critical phenomenological perspective, and when such apparently disparate facets of experience come together, they transform the experience of coding science into one in which the learners in the margins can see themselves as *authors* of code and science, in ways that challenge disciplinary hegemonies.

Aesthetic experience, in the critical phenomenological perspective that we have laid out here, is an account of recentering the margins. It is then a celebration of femininity over masculinity, of caring and togetherness over solipsism and individualism, of heterogeneity over homogeneity, and of voicings of the disciplines' others over their silence in the margins throttled by the patriarchy of the canons. It is the story of how computational objects become meaningful, but not merely personally—to the self and to the others who care for and with each other, and in the embodied and material worlds—without necessarily being folded onto the computational artifact.

Notes

1. J. Dewey (1934). *Art as experience*. Penguin.
2. J. Dewey (1916). *Democracy and education: An introduction to the philosophy of education*, p. 147. Macmillan.
3. P. Sengupta, M. C. Shanahan, & B. Kim (2019). Reimagining STEM education: Critical, transdisciplinary, and embodied approaches. In *Critical, Transdisciplinary and Embodied Approaches in STEM Education*, 3–19. Springer. See also: M. A. Takeuchi, P. Sengupta, M. C. Shanahan, J. D. Adams, & M. Hachem (2020). Transdisciplinarity in STEM education: A critical review. *Studies in Science Education*. doi: 10.1080/03057267.2020.1755802.
4. Dewey, *Art as experience*, 1934.
5. J. Dewey (1950). Aesthetic experience as a primary phase and as an artistic development. *Journal of Aesthetics and Art Criticism*, 9(1), 56–58.
6. A. V. Farris & P. Sengupta (2016). Democratizing children's computation: Learning computational science as aesthetic experience. *Educational Theory*, 66 (1–2), 279–296.
7. P. Kosso (2002). Einstein and the most beautiful theories in physics. *International Studies in the Philosophy of Science*, 16(1), 39–48. See also: G. Engler (2002). The omniscienter: Beauty and scientific understanding. *International Studies in the Philosophy of Science*, 16(1), 27–38.
8. S. Chandrasekhar (1987). *Truth and beauty: Aesthetics and motivations in science*, 64–73. University of Chicago Press.
See also: A. Reuger (2002). Aesthetic appreciation of experiments: The case of 18th-century mimetic experiments. *International Studies in the Philosophy of Science* 16 (1), 51–53.
9. Farris & Sengupta, Democratizing children's computation, 2016, 283.

10. J. Margolis & A. Fisher (2002). *Unlocking the clubhouse: Women in computing.* MIT Press. Margolis, J. (2010).*Stuck in the shallow end: Education, race, and computing.* MIT Press.

11. M. Bakhtin (1990). *Art and answerability.* University of Texas at Austin Press.

12. M. Holquist, page xxx, in M. Bakhtin (1990), *Art and answerability.* University of Texas at Austin Press.

13. Holquist, *Art and answerability,* 1990, *xxx.*

14. A particularly resonant example can be found in the work of undergraduate art students at RISD, who spend an entire year learning experimental physics techniques and instrumentation so that they can create interactive art exhibits by combining everyday objects and scientific objects and theories. See: J. Maeda (2013). *The art of critical making: Rhode Island School of Design on creative practice.* Wiley.

15. M. Bakhtin (1986). *Speech genres and other essays,* 119, University of Texas Press.

16. J. Derrida (2001). *Writing and difference.* Routledge.

17. C. Higgins (2008). Instrumentalism and the clichés of aesthetic education: A Deweyan corrective. *Education & Culture,* 24(1), 7–20.

18. J. Margolis (2010). *Stuck in the shallow end: Education, race, and computing.* MIT Press.

19. L. M. Terman (1923). *Report of sub-committee of committee on scholarship on student ability,* 28. Stanford University Press.

20. Margolis, *Stuck in the shallow end,* 2010, 42.

21. L. Z. Jaber & D. Hammer (2016). Learning to feel like a scientist. *Science Education,* 100(2), 189–220.

22. M. A. Takeuchi (2016). Friendships and group work in linguistically diverse mathematics classrooms: Opportunities to learn for English language learners. *Journal of the Learning Sciences,* 25(3), 411–437.

23. E. R. Sohr, A. Gupta, & A. Elby (2018). Taking an escape hatch: Managing tension in group discourse. *Science Education,* 102(5), 883–916.

24. J. K. Fletcher (2001). *Disappearing acts: Gender, power, and relational practice at work.* MIT Press.

25. M. Anderson. (2018). The radical self-reliance of black homeschooling. *The Atlantic.* https://www.theatlantic.com/education/archive/2018/05/black-homeschooling/560636/

26. K. M. Leander (2002). Locating Latanya: The situated production of identity artifacts in classroom interaction. *Research in the Teaching of English,* 37(Part 2), 198–250.

27. Leander, Locating Latanya, 2002, p. 199.

28. M. G. Ames, 2019. *The charisma machine: The life, death, and legacy of One Laptop per Child.* MIT Press.

29. K. M. Leander & G. Boldt (2018). Design, desire, and difference. *Theory Into Practice,* 57(1), 29–37.

30. J. N. Hale (2016). *The freedom schools: Student activists in the Mississippi civil rights movement.* Columbia University Press.

31. M. A. Takeuchi. (2018). Power and identity in immigrant parents' involvement in early years mathematics learning. *Educational Studies in Mathematics,* 97(1), 39–53.

32. Massumi, *Parables,* 2002, 231.

33. P. Sengupta & A. V. Farris (2012, June). Learning kinematics in elementary grades using agent-based computational modeling: A visual programming-based approach. In *Proceedings of the 11th International Conference on Interaction, Design and Children,* 77–87. ACM.

34. Farris & Sengupta, Democratizing children's computation, 2016.

35. For a detailed discussion of another episode of such aesthetic experience, please refer to: A. V. Farris & P. Sengupta (2016). Democratizing children's computation: Learning computational science as aesthetic experience. *Educational Theory,* 66(1–2), 279–296.

36. J. K. Fletcher (2001). *Disappearing acts: Gender, power, and relational practice at work.* MIT Press.

37. L. McMahon (2017). Phenomenology as first-order perception: Speech, vision, and reflection in Merleau-Ponty. In *Perception and its development in Merleau-Ponty's philosophy,* edited by Kirsten Jacobson and John Russon. University of Toronto Press.

38. V. Shklovsky (1917). Art as technique. In *Art in theory, 1900–1990*, ed. Charles Harrison and Paul Wood. (Blackwell, 1992), 275.

39. Higgins, Instrumentalism, 2008.

40. Higgins, Instrumentalism, 2008, 12.

41. M. G. Ames (2019). *The charisma machine: The life, death, and legacy of One Laptop per Child*. MIT Press.

42. L. Irani (2019). *Chasing innovation: Making entrepreneurial citizens in modern India*. Princeton University Press.

43. Higgins, Instrumentalism, 2008, 10.

44. McMahon, Phenomenology, 2017.

45. Terman, *Report of sub-committee of committee on scholarship on student ability*, 1923, 28.

46. J. Margolis, J. J. Ryoo, C. D. Sandoval, C. Lee, J. Goode, & G. Chapman (2012). Beyond access: Broadening participation in high school computer science. *ACM Inroads*, 3(4), 72–78.

47. Fletcher, *Disappearing acts*, 2008.

48. Ames, *The charisma machine*, 2019, 73–108.

49. Ames, *The charisma machine*, 2019, 131.

50. Ames, *The charisma machine*, 2019.

51. Y. B. Kafai, D. A. Fields, D. A. Lui, J. T. Walker, M. S. Shaw, G. Jayathirtha et al. (2019, February). Stitching the loop with electronic textiles: Promoting equity in high school students' competencies and perceptions of computer science. In *Proceedings of the 50th ACM Technical Symposium on Computer Science Education*, 1176–1182.

52. Ames, *The charisma machine*, 2019, 131.

53. B. Massumi (2002). *Parables for the virtual: Movement, affect, sensation*. Duke University Press, p. 224.

8 Computational Heterogeneity: A Radical Reflection

8.1 Computational Heterogeneity: A View beyond Binaries

But computation—and technology, more broadly—affords a much wider range of experience and possibility than suggested by the box on one's desk.
—Mike Eisenberg[1]

In *Experience and Education*, Dewey argued that humankind is "given to formulating its beliefs in terms of Either-Ors, between which it recognizes no intermediate possibilities." Not surprisingly, educational computing is no exception. One such Either-Or binary that is relevant to our project so far is *computing vs. computational thinking*.[2] A keynote address delivered recently at a computing education conference by the computer scientist and educator Judy Robertson is quite illuminating in this respect.[3] Robertson argued against a blind emphasis on computational "thinking" as the centerpiece of computational literacy. Focusing only on creating SCRATCH programs may lead especially young learners away from understanding how the computer actually works as a "machine." Children's drawings of what is inside a computer show that learning to use computational abstractions that we typically introduce in the K–12 levels—such as variables and loops—does not necessarily help us understand how these abstractions are processed and represented inside the computer. What gets lost in our push to teach children how to code is the *physicality* of computing as a machine, which involves a deeper understanding of the relationship between the hardware and software—and many other forms of computational abstractions that are not even presented to children as part of their experience of coding or computing.

As poignant as Robertson's argument is, we need to be careful before creating and popularizing such binary oppositions in educational computing. Thinking computationally is not merely a solipsistic event, and neither should it imply the reduction of computing and computers to programming and software. Similarly, focusing on computational participation[4] should not negate what can be gained from paying attention to intuitions. The disconnectedness implied in such binaries—hardware versus software, individual versus collective, thinking versus doing, and so on—continues to reify device-level engagement. Within any of these binaries, a shift in focus from one to another simply implies a different form of device-level engagement, for example, a shift from programming to hardware, or a shift from an individual's code to remixed and shared code.

A contrasting image involves fluidity and movement both within and across these binaries. To this end, we offer *computational heterogeneity* as a heteroglossic imagination of code and coding. Committed to accounting for the manifold forms in which code is experienced, it offers an ontological and epistemological reimagination of the boundaries of code. It reminds us of the inseparability of code and its inevitable *other*: the world in which it becomes meaningful in experience, and the voices that are often left out of the folds of computing and STEM.

Computational heterogeneity thus explicitly seeks to counter the scepter of technocentrism, the fallacy of seeking all answers to questions about technology within the technology itself. As we discussed in chapter 1 (see section 1.1), the dangers of technocentrism are two-fold. In the technocentric paradigm, the lack of cultural and institutional supports for computing in educational settings are often treated as a failure of the technology itself,[5] and furthermore, ignoring such factors can lead to positioning historically marginalized students as deficient in computing.[6] In countering technocentrism, computational heterogeneity also involves challenging disciplinary masculinities and hegemonies that have long defined computing and technoscientific cultures.

This implies a shift away from computational artifacts to computational utterances, our central unit of analysis. The forces of centralization and decentralization—that is, the search for coherence as well as the heteroglossic nature that is fundamental to language[7]—both shape computational utterances. It is through the dynamic interplay between these two forces that the experience of coding in STEM shapes the boundaries between code and its *other*—the *social horizon*[8] toward which every utterance is oriented. The separation between what is inside the computer and what is outside becomes malleable, as computer code becomes enmeshed with and refracts through the embodied, material, historical, affective, and social dimensions of experience. In the rest of this chapter, we reflectively examine the epistemological, methodological, and pragmatic significance of the various images of computational heterogeneity that we have presented so far.

8.2 Advancing Critical Phenomenology

As phenomenologists, we positioned our approach in terms of Merleau-Ponty's notion of the *phenomenal field*, centering our focus on the *sense experiences* of students and teachers. The phenomenological agenda also relies on radical reflection, a form of reflection that must necessarily go beyond the veneer of the sphere of givenness.[9] This means searching for sense experiences of the participants in our studies using lenses that go beyond the scope of "computational abstractions" and "computational thinking" and illustrating how these experiences extend way beyond device-level engagements. They are distributed in myriad forms beyond the computer and the programming language and are also constituted through the interactions between them.

As *critical* phenomenologists, we have presented several cases that highlight the importance of *de-centering*[10] the scholarly conversations around code and coding in K–12 STEM education. Experience, in this perspective, is not universal; rather some ways of feeling and perceiving are privileged whereas others are silenced or excluded.[11] Critical phenomenology should always remind us of the threats of masculinity in technology[12] and the historical marginalization and exclusion of gender minorities and racialized voices from scientific and technological spheres.[13]

This implies that we must learn to see privilege and marginalization as inseparable elements of the social horizon toward which a computational utterance is oriented. The metaphors of voicing and heterogeneity must therefore also confront these issues as *central* to framing code and coding, both in general and in K–12 STEM. Specifically, our work outlines several ways to do so: by centering the work of coding in STEM in students' lifeworlds, by centering voices that have been historically left out from the fields of science and technology, by reconceptualizing disciplinary boundaries and using computational simulations to model and discuss social inequalities, and by explicitly recognizing emotion and carework as central to computing in the science or STEM classroom.

Of particular note is the importance of listening to and including teachers' voices in our accounts of coding in K–12 classrooms. This is also an essential element of our critical phenomenological agenda, because teachers' voices remain largely marginalized and are usually excluded in educational computing research. The general approach in educational computing research has been to impose researchers' designs on teachers, and even remove teachers from their classrooms and instead have researchers conduct "intervention" studies. We believe that in order to step beyond technocentrism and technological determinism in the classroom, we must engage with teachers as partners, who in our experience are often women with no prior experience in coding. Chapters 4 and 6 present images of teachers voicing code as an integral part of their mathematics and science classrooms. As we have argued elsewhere, the conceptual dissonance between researchers and teachers that arises in such contexts is central to the heterogeneity that is essential in viewing coding in classrooms as dialogical.[14] The meaning of words and practices that appear disciplinarily profound to reseachers must be negotiated with teachers' perspectives. As researchers, we must remember that the discipline, in a deeply Bakhtinian sense, is only half ours, as it is populated with intentions of others; so, we must learn to see teachers as our essential and inseparable others in reimagining computational science in the K–12 classroom.

Critical phenomenology is thus at once epistemological, methodological, and axiologial. Epistemologically, it is an attempt to reorient our understanding of the experience of computing and coding toward critical, historical, and aesthetic positionings of the participants, away from a view in which coding is tied to device-level engagement. Methodologically, it emphasizes *listening to the heterogeneity in participants' voices* rather than an overt focus on prescriptions of what they should be doing with the computer. Axiologically, there is a distinct commitment against individualism and technocentrism that has largely defined professional cultures of computing and computing education, as illustrated in the work of critical technology scholars such as Noble, Ames, and Irani (section 1.3.2).

A critical phenomenology of code should serve as a caution for us not to rush into designing assessments of coding without paying attention to (1) the complexity and heterogeneity of experiences of coding beyond device-level engagements, and (2) "who" is(are) voicing code, "with" whom, and "for" whom. In urging us to shift from computational artifacts to computational utterances, our goal is to remind ourselves of the difference between the reified text and the heterogeneity inherent in language. This implies a shift from focusing on the computational object as the locus of experience to the ever-changing boundaries between the computer and the more-than-human world as a more complex but authentic site in which coding unfolds.

This is an image of *différance*[15] in which the experience of coding is different and deferred from the symbolic form of code that is given to the learner. It also stands in contrast to a form of authoritarian discourse in which students' computational experiences are primarily evaluated in terms of their computational productions guided by narrow definitions of computational thinking. Computational heterogeneity is an account of how meanings of code and coding emerge through heteroglossic refraction through the alterity and addressivity, perspectivity, and transparency inherent in language, both natural and artificial. In the following section, we present six forms of computational heterogeneity that emerged in our empirical observations and analyses.

8.3 Forms of Computational Heterogeneity

8.3.1 Perspectival Heterogeneity in Coding Science

The first example of a radical reflection on coding is presented in chapter 3, which calls into question a near-axiomatic assumption in the scholarship on agent-based educational computing: taking on the agent perspective is intuitive even for young learners.[16] Chapter 3 suggests that although this may still be generally true, this assumption or claim of intuitiveness of the agent perspective can be seen as an approximation of a much more complex experience when we "zoom in" on the very early moments of how new learners struggle with interpreting agent-based code.

This complexity is revealed in the form of Arnav and Liam's struggles with figuring out what the computational agent stands for (or, should stand for) in their activity. *Thinking like the agent*—which has been shown to be highly effective for helping us understand complex mathematical and scientific phenomena[17]—is by no means itself an easy feat. Multiple perspectival frames or points of view are at work when coding serves as a language for modeling a scientific phenomenon. The heterogeneity arises from perspectives inherent in spoken (natural) language, visuospatial perspectives that are often tied to particular representational formats (e.g., graphs of motion or spatial representations of motion),[18] and then, of course, the perspectives that are prompted by the interpretations of the programming commands. Coding science is therefore a complex language game of a dynamic interaction between perspectives. Add to this the reality of the science classroom, and it becomes even more complex as learners begin to work with others, and teachers intervene. Although "seeing" the world through the perspective of an agent is certainly an important pedagogical goal, paying attention to the "voices" reveals how the agent perspective becomes meaningful only through negotiation with others, even during the early stages of modeling.

What, then, are some implications of perspectival heterogeneity for educational computing researchers? At the very outset, we must be mindful that students' difficulties in understanding and interpreting code, especially in the context of STEM classrooms, may go far deeper than understanding the syntax and semantics of code. It may be that students' challenges arise from the lack of perspectival coherence among the heterogeneous points of view inherent in the different elements of the activity: the phenomena to be modeled, the programming commands to be used, the perspectival complexity of spoken language, the differences in perspectival frames used by different students, and the genre of programming itself (e.g., agent-based programming may present different demands on perspectival framing from other forms of programming such as logic programming). One of the roles that the teacher can play in such settings, as we have illustrated, is to listen carefully for perspectival incoherences and introduce prompts for perspectival shifts accordingly. And finally, we have also pointed out how the modeling platform and activities can be designed in order to support learners to take on multiple and complementary points of view, for example, through supporting both event-based, spatialized representations and graph-based, cumulative representations of motion.

8.3.2   Experiencing Abstractions as Recontextualization

The experience of abstractions, as we remarked in chapter 1, is that of recontextualization. Each of our empirical chapters bears evidence to this effect. The perspectival heterogeneity Arnav and Liam experienced (chapter 3) in order to interpret what a computational agent stands for is an example of recontextualization, in which the complexity of linguistic context shapes the interpretation of code. Our analysis of how new advances in computational modeling of ethnocentrism have emerged as computer scientists have recontextualized and adapted the same algorithm in different phenomenal contexts is an example of another form of recontextualization (chapter 5). The involvement of users within the design process is also an experience of recontextualization of both disciplinary (e.g., mathematical) and computational abstractions, as evident in the work of 4th-grade students described in chapter 4. Teachers, with no prior experience in coding, recontextualize coding as mathematizing in the elementary science classroom, and in the process, they integrate coding, mathematical modeling, and science (chapter 6). In chapter 7, we showed that students in the margins of computing—historically in terms of both their race and their interests—can find their computational voices as their coding brings together different worlds. The virtual world within the computer merges with their lifeworlds outside the classroom. In such contexts, coding becomes carework, the kind that has historically been silenced in the overtly masculine culture of technology design.[19]

It is important to note that this is not a simplistic call to pay attention to the context in which coding is experienced. Very few of our readers would disagree with the general observation that context is central to learning, teaching, and cognition. So, what does our specific observation add here? The most important implication is that the phenomenological frames that we have provided in each chapter—perspectival thinking, transitional othering, designing usable software *for* and *with* others, coding as aesthetic experience, and coding as mathematizing—alert us to specific, heterogeneous *forms* in which coding is experienced in STEM classrooms. This means that our experience of computational abstractions involves the interplay between multiple contexts, and each of the phenomenological frames illustrates a particular form of this interplay. It is this interplay that we term recontextualization. Furthermore, in each form of recontextualization, the relevant experience significantly expands beyond the scope of virtuality on the computer, which in turn necessitates paying careful attention to discourse around computational abstractions, not merely the computational artifacts that embody these abstractions.

8.3.3   Alterity and Addressivity in Computational Design

The notion of disciplinary authenticity runs deep in the scholarship of public education because it determines to a large extent what we want students to know and be able to do, and how we assess learning.[20] What counts as authentic, in turn, is based on our understanding of what disciplinary expertise looks like.[21] What does this mean in terms of educational computing, particularly in the context of coding and modeling in K–12 STEM? One obvious answer is to pay attention to the representational practices that are central to scientific modeling and are also reflexive with programming and computational modeling.[22] Design and perspectival work, as we explained in chapter 2, are examples of two such anchors that can productively pivot us between these worlds.

The emphasis on voicing and heterogeneity brings to light what we believe to be a deeper form of experience underlying authenticity: alterity, or otherness. To remind the reader, alterity is central in Bakhtin's problematizing of the relationship between the "outside" and "inside" of a text, between its origin, context, or referent, and its form or structure.[23] Alterity demands that we pay attention to not only "where" the words are or have come from, but also "whose" words they are. Populated with the intentions of others, according to Bakhtin, language itself occupies the boundary between self and other. A word is thus half ours and half someone else's, and as we argued in chapter 1, what is true of the word must also be true of code.

In chapter 4, we have presented a pedagogical approach in which the "inside" and "outside" of code were reimagined through connecting code with physical models, as well as through actively engaging an authentic audience (mathematics teachers). The experience of the *other*—the authentic audience—took on a central role in the students' design journey. Mathematics teachers unaffiliated with the study served as prospective users of the mathematical machines designed by student dyads and were involved as an integral part of the design experience. What counts as mathematical explanations of code became redefined for the students through iterative engagement with the users, which also deepened their engagement with the code.

In Bakhtinian parlance, this is also an experience of addressivity, the constant state of being addressed and being in the process of answering. Paying attention to addressivity implies a shift away from computational artifacts to computational utterances as the focus of computational design. Enlivening addressivity involved moving beyond the technocentric frame in the classroom as the student-designers iteratively interacted with authentic users in two successive user testings. Their teacher, Ms. Lena, also carefully designed classroom instruction through which she helped students recognize the users' feedback as valuable for improving their designs. It was through such experiences that student-designers learned to see their own work as designs *for others*. Through this work, we have also critiqued the individualistic emphasis in computing education. Notions of agency, ownership, and control are repositioned dialogically, as students' computational designs embody iterative improvements that arose from dialogical engagements with each other and the users.

8.3.4   Mathematizing: Computational Heterogeneity and Teacher Voice

In chapter 6, we followed an elementary teacher's (Emma's) class over a span of two years, as she used programming with ViMAP as part of her science and math classroom instruction on a weekly basis. Emma framed coding as "mathematizing," that is, she found a place for coding in her science classroom as a way to mathematically model physical and biological phenomena that students were learning. Within this broader frame of mathematizing, the most persistent finding was the constant movement of the modeling activities across representational systems: embodied modeling, physical modeling, modeling using programming commands in ViMAP, and modeling using mathematical inscriptions (e.g., multiplication tables, hand-drawn geometric shapes, and so forth). Coding took on the form of a phenomenon that circulated across these representations, which involved transforming one form of representation to another.

Integrating coding with K–12 science is not merely a problem of designing programming languages that are aligned with disciplinary expectations; we must also keep in mind the inherent diversity of the representational infrastructure within the classroom that teachers typically work within, alongside the programming language. It is in working with the non-computational elements of this infrastructure that teachers may be able to find their computational voices, as they engage in the work of designing modeling activities that involve both programming on the computer and modeling outside the computer. The ruptures that appear at the interfaces between heterogeneous representational forms also become sites for productive inquiry.

By leveraging and making visible the heterogeneity of the representational infrastructure, Emma created opportunities for children by valuing, rather than devaluing, their interpretive dilemmas and moves, and by acknowledging the uncertainties involved in managing the *mangle*[24] of materiality, theorization, and coding in the context of modeling science.

8.3.5 Relational Work: A Critical Aesthetics of Coding

The roots of Logo in feminist epistemology have been well documented in early writings on constructionism. These writings reflect how learning with a computational agent (e.g., a Logo Turtle) is more akin to getting to know a person, rather than a mastery of symbolic forms, and thus highlights a relational epistemology that stands in contrast to technological determinism (see section 1.3.2). But is simply using Logo or agent-based computing (more broadly) enough to break the masculinist hegemony in technological worlds?

Ames' groundbreaking observations of the context of how the OLPC laptops and software were *actually* used by the intended users—children in the Global South (for example, Paraguay)—is a reminder of the limits of such technocentric imaginations, in which the essence of learning is ascribed to the "charisma" of technology. She illustrates the richness of childrens' lives away from the computer and contrasts that with the limited experiences on the computer.[25] In a similar sense, chapter 7 highlights how important the performances and relationships outside and beyond the computer and the classroom are for learners who are at the margins of computing for finding themselves within computing.

Papert's notion of the Turtle as a transitional object is a certainly powerful proposal. The ViMAP Turtle "stood in" for a religious symbol for Shenice, and for Ariana's best friend: Thomas. But caring for Thomas also brought Ariana together with Matt, another student in the course, and created opportunities for humorous banter between them. Ascribing the richness of Ariana and Matt's relationship to the power of the transitional object, however, would be essentially technocentric. The richness of the experiences that unfolded outside the computer brings into account the lifeworld of the learners, including who (not only what) they care about, and even their religious lives, as was evident in Shenice's case.

So, to answer the question we asked in the first paragraph: no, simply using Logo-like programming environments is not the answer. But designing activity systems in which children, especially those who have been historically marginalized due to their race and gender, can meaningfully integrate stories of their loved ones and their objects of affection with disciplinary work, can offer a critical and aesthetic[26] alternative to technocentrism. The focus here is on a deep and inextricable relationship between *history* and *form*, which Bakhtin argued is inextricable in language.[27] The historical dimension of the learners' experiences becomes deeply intertwined with their computational work as they engage in a particular *form* of experience: relational work.

A critical aesthetics of coding thus orients our attention to relationality in learners' experiences beyond the computing device(s), not merely its structural elements. This also includes the personal, social, and historical dimensions of their experiences, as evident in our account of Shenice's work in chapter 7. Caring for the other—where the other stands in for the computational agent as well as friends and loved ones—is an example of such a form of experience that at once is historically grounded in the lives of the learners and also makes explicit a relational epistemology of computing. But, our point here is that this critical, aesthetic manifestation of computing as language becomes visible when we, as researchers and educators, learn to see the happenings beyond the device as the spaces where computing also unfolds. While Ariana and Matt's banter offers a fortuitous example of such a relational unfolding, our inability to recognize and honor Shenice and her friends' invitation to their church stands as an example of our device-centeredness. The charisma of the computer can—and typically does—blind us to such relational unfoldings. Given that relational work is actively discouraged and "disappeared" from the professional worlds of technology design,[28] it becomes even more important for us to value such *forms* of experience in the computing classroom.

8.3.6   Voicing Code as Transitional Othering and Recontextualization

How can computational modeling be used to support critical and difficult conversations about race and inequality in the classroom? In chapter 5, we take on this issue and argue that *transitional othering* can offer us a potential answer. The transitional nature of computational agents can provide students and teachers with a space in which they can explore complex critical sociopolitical issues such as race and economic inequalities by drawing upon their personal experiences, albeit in a manner that allows them to maintain a distance from their personal lives. This is similar to Mead's notion of "distance experience," in which people move "beyond themselves" into the future, constructing imagined worlds for escaping the circularity of social reproduction of racial oppression. The computational agent becomes a projection of the "self," as well as a representation of the "other," whom students would normally not identify with themselves in their daily experiences.

However, it is important to note that our goal is not to position the computational agent as a panacea. The instructional context for supporting experiences of transitional othering involved recontextualizing an algorithmically generated multiagent simulation of ethnocentrism in light of the Racial Dot Map. We also provided an example from the field of computational social science that highlights the importance of progressive recontextualization of generalized computational abstractions in advancing scholarship on computational modeling of ethnocentric phenomena. In our classroom study, going beyond the technocentric imagination, we provided phenomenological accounts in terms of the lived experiences that the different participants brought into play in their discussions, once they were able to recontextualize the ethnocentrism simulation in terms of the Racial Dot Map. It is through the experience of recontextualizing that preservice teachers were able to dive deeper into their explanations and understandings of the simulation as a representation of racial and socioeconomic inequalities. The computational abstractions used in the ViMAP simulation (e.g., the agent-level variables) became recontextualized as attributes of and interactions between people represented by the dots in the Racial Dot Map.[29]

Transitional othering is an act of *ostranoniye*:[30] making the familiar strange, and perhaps more importantly, making the strange familiar. A form of ostranoniye that is particularly relevant to our work is what critical race scholars have termed *projectivity*—an imaginative generation of possible alternate trajectories of action by racialized actors.[31] Projectivity allows racialized actors to distance themselves from and challenge problematic habits and traditions. It also creates opportunities to reformulate these habits and traditions that otherwise perpetuate racial oppression, marginalization, or silence around these issues (e.g., in our case, the culture of silence around topics that involve the Civil War in classrooms in the Southern US). However, it is not only racialized people who can benefit from such opportunities. As we show in chapter 5, transitional othering enables a nearly all-White class of preservice teachers to engage in critical conversations and reasoning about the complex relationships of race and socioeconomic inequalities in the US. This is a significant finding, given that it has been shown to be challenging for White teachers to engage in such conversations in their K–12 classrooms, even when they are teaching social studies.[32]

8.4   Epilogue: Lessons for Avoiding Technocentrism

In concluding our book, we return to Papert's call to take the literary metaphor seriously in order to develop a culture of "computer criticism." In going beyond a structuralist analysis of language and literature, literary criticism takes on a problem that is also at the heart of phenomenology: *enframing*.[33] The central purpose of literary criticism is to highlight the complex relationships between what is included within such enframings and what is excluded from them. In Heideggerean terms, the frame itself becomes the meaning of language. What happens when we take the linguistic metaphor seriously in the context of educational computing? Our concern here is to avoid coining aesthetic clichés[34]—that is, phrases that have only rhetorical power—rather than actually orienting our actions toward expansive imaginations of coding. Can computational heterogeneity actually reorient our praxis as researchers?

Epistemologically, this book in its entirety is an argument against a particular form of enframing of the experience of coding that relies only or primarily on device-level engagements, typically amplifying our focus on the structural elements of code. Wing's call for computational thinking—however myopic its interpretation and uptake by the educational computing community has been— was fundamentally an attempt to orient our attention to the experiential nature of computing and coding. However, the discourse around computational thinking is largely dominated by a rather superficial conceptualization of computational abstractions, typically represented in terms of structural elements within the programming language. Such an enframing creates divisions that often might not reveal the underlying complexity of the relationships between elements "inside" and "outside" the frame. This is essentially the problem of technocentrism and technological determinism: elements within the technology become primary or sole representations of our experience. In this concluding section of our book, we reflect on some lessons from the field that we have learned over the last decade of *listening* to the voices of hundreds of students and teachers that can help us look beyond technocentrism and avoid technological determinism in our inquiry.

8.4.1   Lesson 1: Worldliness Beyond Microworlds (and Microcontrollers)

In the classroom, we give accounts of the world beyond.
—Gayatri Chakravorty Spivak[35]

The worldliness of the computer, since Papert's time, has been rooted in the malleability of computing software and hardware. From the shapeshifting versatility of Proteus, the Greek god, comes the metaphor *protean*, which Papert used to describe the computer's power. Wing's notions of computational abstractions and computational thinking provide another way of conceptualizing the protean nature of computing through the *automation of abstractions*.[36] The "capital" of computing is located in the work of representing the world in a microcosm—microworlds—which, while powerful in a protean sense, can also portray an image of "purity, simplicity, and seclusion."[37]

Voicing code fundamentally orients us to the worldliness of code beyond microworlds and the microcontroller, even when every chapter in this book involves participants working with microworlds. The images of coding we have presented illustrate how code is voiced through and alongside heterogeneous elements of students' and teachers' lived experiences: their religious lives, designing not only for but *with* others, carework and personal attachments to friends and family members, conversations about race through transitionally othering code, and teachers' reframings of coding as mathematizing. The meaning of code and coding thus becomes significantly amplified, and perhaps more importantly, *transformed*, as students, preservice teachers, and teachers find their computational voices.

The worldliness of code, for educational designers and researchers, must go beyond representing the world inside the microworld. It must untether our imaginations of code and coding from device-level engagements. The transparency of code and computational agents that can be leveraged to represent urban segregation in a microworld must be complemented by attention to transparency afforded through pedagogical talk. And while microcontrollers like Arduinos can expand computing by enfolding materiality—for example, distributed mathematical machines discussed in chapter 5—it is through enlivening the addressivity in discourse that computational utterances find their social horizons. The alterity inherent in designing a computational artifact must be complemented by *being with* users.

And finally, it is important to remind ourselves that our call for worldliness should not blind ourselves to the oppressive and hegemonic histories and ideologies in which STEM disciplines and computing are entrenched.[38] The world that is revealed to us through a critical phenomenology of code must actively work toward making the professional and pedagogical worlds of computing and STEM less oppressive by centering voices from the margins. For example, we must be mindful of the fact that racialized, immigrant labor constitutes much of the computing industry in the US, and developing a deep understanding of the intersectionalities of gender, race, and migration must also inform our understandings of what counts as authenticity in computing education.[39] Authenticity in K–12 computing and STEM education has also been critiqued due to the cisheteronormativity inherent in these fields of practice, which in its assumed naturalness of gender binaries and heteronormative relationships, has systematically and historically ignored, silenced, and oppressed people who identify as queer and non-binary.[40] Another form of critique of authenticity involves decolonizing computing and complexity education in partnership with Indigenous communities, which can also alter our commonly accepted, Westernized notions of programming and modeling using digital programming languages.[41]

While deeper explorations of these specific issues are beyond the scope of our book, by centering worldliness, we essentially argue that a commitment to centering systemically and historically marginalized voices is urgently needed for the field. This, in turn, necessarily involves paying attention to heterogeneity, and the metaphors that we have presented in this book may be helpful. But more importantly, critical phenomenology is a reminder that what we know as coding may simply be a second-order perception that stands to be changed as we listen carefully to voices from the margins. Worldliness, thus, is not merely a call for orienting our attention to the place where code unfolds; it may necessitate changing the world of code itself.[42] It is a commitment to carrying "the burden of recognition"[43] of the injustices that masculine, individualistic, and militaristic disciplinarities of technoscience have historically and systematically enacted on Black people, Indigenous people, women, people of color, and queer and non-binary people, while simultaneously working toward learning to be in solidarity with them.[44]

8.4.2   Lesson 2: Data as Trouble

The technocentric universality[45] implied in viewing computing as the automation of computational abstractions can stand in contrast to the heterogeneity inherent in modeling science. Whereas algorithms and data structures can be used effectively and reflexively in scientific work, engaging with the material world leads to "trouble": kinks and fractures in the assumed infrastructure that constitutes disciplinary work, and ruptures in the seamlessness implied in the automation of abstractions. The automaticity of algorithmic computations of distance traveled by the ViMAP Turtle on one hand, and the messiness and uncertainty involved in students' measurement of step sizes on the other, is an illustration of this contrast (chapter 6). What emerges then is the value of bringing together both these forms of experience within the fold of computational modeling. Furthermore, as Emma's classroom in Year 2 illustrates, teachers can come to value this uncertainty as a site for conceptual and representational innovations in modeling.[46]

As we have noted elsewhere, Dewey's critique of empiricist ontology is similar.[47] Dewey argued that the quest for certainty is the hallmark of modernity and positioned empiricist ontology at the root of this quest.[48] He critiqued empiricist ontology because it replaces the emergent nature of experience with data. Whereas empiricism often represents data as self-evident, Dewey positions data as signifying a phenomenon for further inquiry. Data, or objects for that matter, then, do not represent things-in-themselves, but "are manifestations of particular kinds of novel and complex relationships that take place in and through time,"[49] which can only be understood through further inquiry. This pragmatist critique of empiricist ontology pre-dates Heidegger's work, but bears a deep similarity in pointing out that when data is taken to be "self-evident," then knowledge becomes an *antecedent reality*—a view that is in direct opposition to the différant nature of experiences of coding we have illustrated in this book. In Heideggerian terms, such an approach *enframes* knowledge by hiding the experience behind the data, or even making it irrelevant. In such cases, in Merleau-Ponty's terms, there is no "rupture with familiarity"; instead, we fall back on the spheres of givenness—the second-order perceptions of technology—that continue to recursively appear as the answers to our questions about technology.

In positioning trouble as central, and even desirable, we can confront the myopic and problematic discourses around grit and persistance,[50] and reframe failure as an essential, recursive, and iterative experience *where* learning happens.[51] This is particularly relevant for working with marginalized students who have been *systematically*[52] labeled as "failures," a label that is hard to overcome.[53] However, it is also important to note that the heterogeneity in experiences of trouble necessarily go beyond the image of the individual learner. The charismatic computer must not be replaced by the individualistic image of child-centeredness that ignores or omits the important role of joint activity. This in turn necessitates paying attention to assistance from teachers (chapter 6), other students (chapter 3), and intended users (chapter 4), and must also foreground, rather than ignore, students' interpretive dilemmas and challenges that are at once epistemic and representational in nature (chapter 6).

8.4.3 Lesson 3: Code as a Boundary Layer

Our version of computer criticism is premised on positioning coding as *voicing*, a heteroglossic lens we borrowed from Bakhtin. Voicing presents an image of heterogeneity that renders fluid the boundary between natural and artificial languages, between what lives inside the computer and what's outside. This is essential because in the absence of such fluidity, the experience of code is folded onto code itself. *Voicing* opens the space up for bringing into account sociological, political-historical, and affective realms of experience in which coding comes alive in discourse. The question of learning to code can then be reframed as an inquiry into how students and teachers develop their computational voices. The forces that shape their voices include both univocality and multivoicedness, and it is the dynamics between these forces that becomes the phenomenon of inquiry for us as researchers and educators. One could argue that recent studies that rely on data mining the online computational artifacts created by students also reveal how computational abstractions are "voiced," for example, as students engage in "remixing" others' code.[54] However, in the absence of richer analyses of the non-computational dimensions of the experience of the learners, the image of learning to code gets folded within code itself.

Computational heterogeneity is a form of simultaneous ideological broadening and deepening—an ongoing attempt to redraw lines around what forms of experience should *also* count as coding, especially in K–12 science and math. Gieryn introduced boundary work as an ideological style in scientists' attempts to create a public image for science by contrasting it favorably to nonscientific intellectual or technical activities. So, descriptions of science as distinctively truthful, useful, objective, or rational may also be interpreted as ideologies that, although incomplete and ambiguous as images of science, are nevertheless useful for "scientists' pursuit of authority and material resources."[55] In a similar sense, we wonder if contestations around defining computational thinking can be characterized as ideological contestations and commitments, rather than matters of scientific verity. For example, by positioning computational thinking as thinking about computational abstractions or by claiming that everyday situations can be better understood in terms of computational abstractions, we reveal a commitment to generalizablity (see section 1.3.1). Similarly, the perspective that computational thinking is inseparable from representational practices reveals a different form of disciplinary commitment that values uncertainty in technoscientific work.[56] Our book is also an attempt to argue for an ideological expansion and deepening by arguing for a heightened emphasis on heterogeneity, noting that the nature of computational experiences is *fundamentally* heterogeneous—particularly in K–12 STEM contexts—and far more heterogeneous than what has been previously reported in the literature.

Gieryn reminds us that science acquired its intellectual authority through boundary work: the public demarcation of "science" from "non-science." Although this book is not the public sphere and, like many other academic books, will live in a realm that requires privileged access, it is nevertheless an effort to redraw lines of separation and mergings between code and human experience— a form of ongoing distinction and difference-making that is fundamental to ideological constructions of both theories of computing and theories of learning. Shanahan's metaphor of *boundary layers*[57] then aptly captures our accounts of coding in K–12 STEM: new places of encounters and interactions, akin to the places where different fluids meet and create a boundary layer. When seen in this light, code can become a space where disciplinary worlds of computational science can coexist alongside socially and personally meaningful interpretations beyond the designer's imaginations. For example, a cluster of flocking computational birds can become a discursive space for conversations about overcrowded prisons between a mother and her child.[58]

The *experience* of code is different and much more complex than code itself. The literary metaphor at the heart of our work also reminds us that coding, like Bakhtin's language, is a quest for words that are not our own, that are différant from code itself, an act of "seeing more" rather than merely "seeing as."[59] It is this very act of seeing more in which we engaged—reimagining how to account for experiences of coding in expansive ways—that we invite the field to engage in, rather than unproblematically adopting preset categories for looking at computing, including our own. The essence of coding is the heterogeneity of experience, not device-centered commitments to technology, and educational computing researchers must always be alive to it.

Notes

1. M. Eisenberg (2003). Mindstuff: Educational technology beyond the computer. *Convergence*, 9(2), 29–53.
2. Another example of such an "Either-Or" framing is computational thinking vs. computational participation. We direct the reader to Kafai and Burke's book *Connected Code*, which explores computational participation in out-of-school settings. Y. B. Kafai & Q. Burke (2014). *Connected code: Why children need to learn programming*. MIT Press.
3. J. Robertson (2018). Answering children's questions about computers. *Communications of the ACM*, 62(1), 8–9.
4. Y. B. Kafai & Q. Burke (2014). *Connected code: Why children need to learn programming*. MIT Press.
5. S. Papert (1987). Information technology and education: Computer criticism vs. technocentric thinking. *Educational Researcher*, 16(1), 22–30.
6. J. Margolis (2010). *Stuck in the shallow end: Education, race, and computing*. MIT Press.
7. T. Todorov (1984). *Mikhail Bakhtin: The dialogical principle*. Manchester University Press.
8. Todorov, *Mikhail Bakhtin*, 1984, 56.
9. L. McMahon (2017). Phenomenology as first-order perception: Speech, vision, and reflection in Merleau-Ponty. In *Perception and its development in Merleau-Ponty's philosophy*, edited by K. Jacobson & J. Russon. University of Toronto Press.
10. P. Banerjee & R. Connell (2018). Gender theory as Southern theory. *Handbook of the sociology of gender*, 57–68. Springer.
See also: B. Hooks (2000). *Feminist theory: From margin to center*. Pluto Press.
11. S. Ahmed (2006). *Queer phenomenology: Orientations, objects, others*. Duke University Press.
12. W. Faulkner (2001, January). The technology question in feminism: A view from feminist technology studies. In *Women's Studies International Forum*, 24(1), 79–95. Pergamon.
13. J. Margolis & A. Fisher (2002). *Unlocking the clubhouse: Women in computing*. MIT Press.
See also: S. Vakil (2018). Ethics, identity, and political vision: Toward a justice-centered approach to equity in computer science education. *Harvard Educational Review*, 88(1), 26–52.
14. A. C. Dickes, A. V. Farris, & P. Sengupta (2020). Sociomathematical norms for integrating coding and computational modeling with elementary science: A dialogical approach. *Journal of Science Education and Technology*, 29, 35–52.
15. J. Derrida (2001). *Writing and difference*. Routledge.

16. For a review, please see: R. L. Goldstone & U. Wilensky (2008). Promoting transfer by grounding complex systems principles. *Journal of the Learning Sciences*, 17(4), 465–516.

17. Even some of our own previous research has shown that the intuitive alignment of the agent perspective with learners' prior experiences makes it possible for even 5th graders to learn complex scientific phenomena that are typically introduced to postsecondary students. See: P. Sengupta & U. Wilensky (2011). Lowering the learning threshold: multiagent-based models and learning electricity. In *Models and modeling*, 141–171. Springer.

18. O. Parnafes (2007). What does "fast" mean? Understanding the physical world through computational representations. *Journal of the Learning Sciences*, 16(3), 415–450.

19. See our discussion of Joyce Fletcher's work in chapter 7.

20. G. Wiggins (1989). A true test: Toward more authentic and equitable assessment. *Phi Delta Kappan*, 70(9), 703–713.

21. See, for example, the "images" of what scientists do that guide how science education has been conceptualized, in R. Lehrer & L. Schauble (2006). Scientific thinking and scientific literacy: Supporting development in learning contexts. In *Handbook of child psychology*, edited by K. A. Renninger & I. Sigel. V. IV. Wiley.

22. See P. Sengupta, J. S. Kinnebrew, S. Basu, G. Biswas, & D. Clark (2013). Integrating computational thinking with K–12 science education using agent-based computation: A theoretical framework. *Education and Information Technologies*, 18(2), 351–380; See also D. Weintrop, E. Beheshti, M. Horn, K. Orton, K. Jona, L. Trouille, & U. Wilensky (2016). Defining computational thinking for mathematics and science classrooms. *Journal of Science Education and Technology*, 25(1), 127–147.

23. D. Carroll (1983). The alterity of discourse: Form, history, and the question of the political in M. M. Bakhtin. *Diacritics*, 13(2), 65–83.

24. See our discussion on Pickering's *Mangle of Practice* in section 6.2. See also: P. Sengupta, A. C. Dickes, & A. V. Farris (2018). Toward a phenomenology of computational thinking in STEM education. In *Computational thinking in the STEM disciplines: Research highlights*, edited by M. S. Khine. Springer.

25. M. G. Ames (2019). *The charisma machine: The life, death, and legacy of One Laptop per Child*. MIT Press.

26. See our detailed discussion of Deweyan aesthetic experience in chapter 7.

27. P. N. Medvedev, M. M. Bakhtin, & A. J. Wehrle (1991). *The formal method in literary scholarship: A critical introduction to sociological poetics*, 60–61. Johns Hopkins University Press.

28. See our discussion of Joyce Fletcher's book *Disappearing acts* in chapter 7: J. K. Fletcher (2001). *Disappearing acts: Gender, power, and relational practice at work*. MIT Press.

29. See for example our discussion of Jamie's work in section 5.3.

30. V. Shklovsky (2015). Art, as device. *Poetics Today*, 36(3), 151–174.

31. M. Emirbayer & M. Desmond (2015). *The racial order*. University of Chicago Press.

32. See chapter 5 for a detailed discussion.

33. J. Culler (2008). *On deconstruction: Theory and criticism after structuralism*. Routledge.

34. C. Higgins (2008). Instrumentalism and the clichés of aesthetic education: A Deweyan corrective. *Education and Culture*, 24(1), 7–20.

35. G. C. Spivak (2012). *An aesthetic education in the era of globalization*, 335–350. Harvard University Press.

36. J. Wing (2017). Computational thinking's influence on research and education for all. *Italian Journal of Educational Technology*, 25(2), 7–14.

37. Eisenberg, Mindstuff, 2003, 39.

38. M. A. Takeuchi, P. Sengupta, M. C. Shanahan, J. D. Adams, & M. Hachem (2020). Transdisciplinarity in STEM education: a critical review. *Studies in Science Education*, 56(2), 213–253.

39. T. M. Philip, & P. Sengupta (2020). Theories of learning as theories of society: A contrapuntal approach to expanding disciplinary authenticity in computing. *Journal of the Learning Sciences*. Available at: https://doi.org/10.1080/10508406.2020.1828089.

40. D. Paré & P. Sengupta. (In Press). Queering computing and STEM education. *Oxford Research Encyclopedia of Education*. Oxford University Press.

41. M. Lam-Herrera, I. A. Council, & P. Sengupta (2019). Decolonizing complexity education: A Mayan perspective. In *Critical, Transdisciplinary and Embodied Approaches in STEM Education*. Springer.

42. See for example: D. Paré, M. C. Shanahan, & P. Sengupta (2020). Queering complexity using multi-agent simulations. In *Interdisciplinarity in the Learning Sciences, 14th International Conference of the Learning Sciences (ICLS)*, 1397–1404. ISLS.

43. Philip & Sengupta, Theories of learning. In press.

44. P. Sengupta (2020). Re-orienting design: An unbearable pain. In *International Conference of the Learning Sciences (ICLS 2020)*. ISLS. Available at: http://hdl.handle.net/1880/112215

45. See section 1.3.1.

46. See also: A. V. Farris, A. C. Dickes, & P. Sengupta (2019). Learning to interpret measurement and motion in fourth grade computational modeling. *Science & Education*, 28(8), 927–956.

47. Sengupta, Dickes, & Farris, Toward a phenomenology, 2018.

48. J. Dewey (1929) 1984. *The later works of John Dewey 1929: The quest for certainty*. SIU Press.

49. A. Stoller (2018). Dewey's creative ontology. *Journal of Thought*, 52(3–4), 47.

50. S. Vossoughi, P. K. Hooper, & M. Escudé, (2016). Making through the lens of culture and power: Toward transformative visions for educational equity. *Harvard Educational Review*, 86(2), 206–232.

51. P. Blikstein (2013). Digital fabrication and "making" in education: The democratization of invention. In J. Walter-Herrmann & C. Buching), *Fablabs: Of machines, makers and inventors*, 1–21. Bielefeld (transcript).

52. Vossoughi, Hooper, & Escudé, Making through the lens, 2016, 216.

53. L. Martin (2015). The promise of the maker movement for education. *Journal of Pre-College Engineering Education Research* (J-PEER), 5(1), 30–39.

54. See for example: S. Dasgupta, W. Hale, A. Monroy-Hernández, & B. M. Hill (2016, February). Remixing as a pathway to computational thinking. In *Proceedings of the 19th ACM Conference on Computer-Supported Cooperative Work & Social Computing*, 1438–1449. ACM.

55. T. F. Gieryn (1983). Boundary-work and the demarcation of science from non-science: Strains and interests in professional ideologies of scientists. *American Sociological Review*, 781–795. Quoted segment can be found on page 793.

56. R. Duschl (2008). Science education in three-part harmony: Balancing conceptual, epistemic, and social learning goals. *Review of Research in Education*, 32(1), 268–291.

57. M. C. Shanahan (2011). Science blogs as boundary layers: Creating and understanding new writer and reader interactions through science blogging. *Journalism*, 12(7), 903–919.

58. P. Sengupta & M. C. Shanahan (2017). Boundary play and pivots in public computation: New directions in STEM education. *International Journal of Engineering Education*, 33(3), 1124–1134.

59. Higgins, Instrumentalism, 2008.

Index